DOOMED, UNLESS

How Climate Change and *Political Correctness* will Destroy Modern Civilization

Luc Gagnon, M.Sc., Ph.D.

Doomed, Unless
Copyright © 2022 by Luc Gagnon, M.Sc., Ph.D.

Illustrations by *Violette Moukhtar*

Tellwell Talent
www.tellwell.ca

ISBN
978-0-2288-8363-0 (Hardcover)
978-0-2288-8362-3 (Paperback)
978-0-2288-8364-7 (eBook)

Image on Book Cover

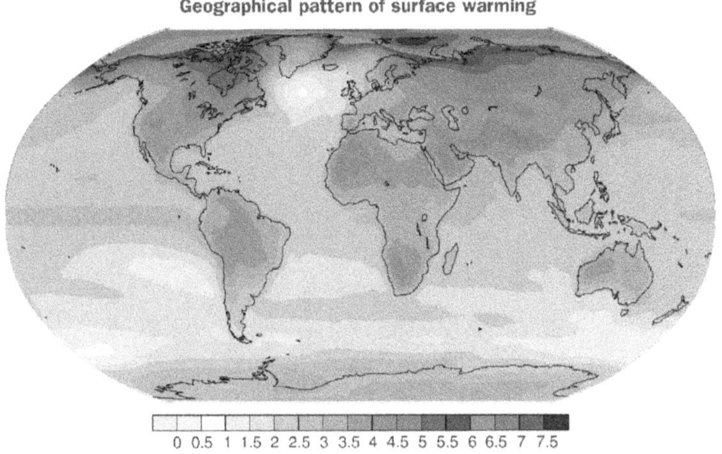

Figure SPM.6. Projected surface temperature changes for the late 21st century (2090-2099). The map shows the multi-AOGCM average projection for the A1B SRES scenario.
Legend in degrees C
Warming of 7.5° C = 13.5° F

IPCC, 2007: Climate Change 2007: Synthesis Report. Contribution of Working Groups I, II and III to the Fourth Assessment Report of the Intergovernmental Panel on Climate Change. IPCC, Geneva, Switzerland, 104 pp.

Table of Contents

Reviews of DOOMED, UNLESS by Energy Experts

Normand Mousseau

- Professor in Physical Science, Montreal University
- Scientific director, Trottier Energy Institute

Luc Gagnon's book is a breath of fresh air in the climate debate. Based on rigorous analysis and a critical eye refined over his long career, the author offers us a raw picture of the failure of climate policies. Too often, these policies are fueled by good thinking and political correctness rather than by facts. The result is billions of dollars spent ineffectively, inadequate citizen efforts, and a planet falling apart. Compared to the disasters ahead, the COVID-19 pandemic will have been negligible. And yet, the response of governments to the pandemic was much more decisive and ruthless, than the actions on climate change. Luc Gagnon allows us to understand the underlying issues better; he identifies courses of action that could change things. A book to read and discuss absolutely.

Pierre-Olivier Pineau

- Professor, Hautes Études Commerciales, Montreal University
- Lead author of the annual report *Energy in Quebec*

We need a new way of understanding the energy transition. This book sheds a harsh light on why inertia is winning, despite the importance of our scientific knowledge. It presents, with extraordinary rationality, the fact that it is not technological barriers that prevent decarbonisation, but a refusal to review our lifestyles. The solutions are within reach, but they require changes at many levels. Almost no one has the courage to clearly identify these changes for fear of offending this or that group. Luc Gagnon has that courage.

Paul Lanoie

- Sustainable Development commissioner, Québec government
- Professor, Hautes Études Commerciales, Montreal University (for 27 years)

Luc Gagnon has undeniable expertise in climate change. He paints a clear, instructive, and insightful picture of the current situation and the huge challenges we face. In doing so, he lucidly tackles taboo subjects such as population growth or GHG emissions from pets. His provocative observations are certainly worth considering, as are the solutions he advocates. In this regard, as a former Commissioner of Canada's Ecofiscal Commission, I can only approve his proposal on ecological tax reform!

Jean-François Boisvert

- President and co-founder, Montreal Climate Coalition

To avoid the climate catastrophe looming on the horizon, we must free ourselves from fossil fuels. Since fossil fuels currently provide 80% of the energy consumed worldwide, this represents an enormous challenge. Luc Gagnon's demonstration is essential on several levels: small gestures will not be sufficient; actions must be much more ambitious and attack the problem at its root; our political leaders, to avoid displeasing their constituents, do not dare to advance on this path. Even environmental organizations prefer to avoid certain subjects. However, it is necessary to be able to debate them as lucidly as possible. The merit of this book is to tackle these issues head-on.

Jean-François Lefebvre

- Lecturer in sustainable development for 14 years, Urban Planning, Quebec University at Montreal

Luc Gagnon offers us a masterful work. He ignores prejudices, shakes up taboos, and goes beyond preconceived ideas. Despite the consensus, progressive circles, on the urgency of tackling the climate crisis, we are far from being "out of the woods". However, the author shows us that the solutions are very tangible, right there in front of us: bold urban redevelopment, ambitious ecological fiscal reforms… He has the lucidity to accept the real causes plaguing humanity and ultimately threatening its survival. Clearly, without deceit. An essential work.

Introduction:
The cycle of empty promises

In the first sixteen months of the COVID-19 pandemic, the virus caused four million deaths worldwide. Is this exceptional when compared to other global issues? In comparison, air pollution from fossil fuels is responsible for 6 or 7 million deaths, **each year**[1]. This is due to pollutants such as nitrogen oxides, particulate matter, and volatile organic compounds. The assessment does not consider greenhouse gas emissions (GHGs), and the effects of climate change.

A complete assessment should include climate change, which increases the intensity and frequency of floods, hurricanes, tornadoes, heat waves, droughts, bushfires, and forest fires. In the future, the annual number of victims will be calculated in millions. A National Geographic article[2] confirms this probable outcome for heat-related impacts:

> *The ultimate solution to global warming, of course, is to drastically reduce our greenhouse gas emissions. If we utterly fail to do that, by 2100, the heat-related death toll could rise above 100 000 a year in the U.S. Elsewhere, the threat is far greater: In India, for example, the death toll could reach 1.5 million.*

The article mentions another probable impact in 30 years:

---

[1]   United Nations Environment Programme, 2019.

[2]   National Geographic, August 2021, p. 48.

> *By 2050, high heat and humidity in the American Southeast will likely render the entire growing season unsafe for agricultural work with present-day working practices.*

In sum, portions of current human habitat will become unfit for life.

Despite this, we will describe, in the first chapter, why international institutions constantly **underestimate** the impacts of climate change. One reason for downplaying the impacts is *political correctness*.

What is *political correctness?*

It is a social code that prohibits language or actions that may offend or disadvantage others. It becomes unacceptable to say that a particular group is creating a problem. For example, it is unacceptable to say that the suburban lifestyle consumes a large quantity of energy.

Yale researcher S. Morris[3] explained the process that causes *political correctness*: politicians care very much about what people think of them; as a result, they change their behavior or opinion, to protect their reputation. This is not surprising in the case of politicians who must be elected. The contribution of Morris is to show how **this process also applies to most experts that give advice to politicians**. If these experts want to benefit from future subsidies or contracts, they need to adopt the expected behavior or conclusions.

---

[3]   Journal of Political Economy, 2001, Political Correctness.

In this context, *political correctness* must include friendly characteristics: be positive, create hope and propose a few symbolic solutions, even if they are simplistic. So, instead of describing the energy consumption of suburbs, politicians and environmentalists promote the recycling of residential waste. Recycling becomes a tool to avoid real issues.

The pattern of broken promises

Managing climate change is an urgent challenge that requires economic and fiscal reforms. Despite this, the following pattern, greatly influenced by *political correctness*, has been repeated for more than 30 years:

- Scientists who study climate change understand the fundamental requirements to reduce emissions. But being dependent on government funding, they do not want to offend politicians. They want to continue as advisors, so they make formal assessments of climate change that are acceptable to politicians, underestimating the impacts.
- In response to these scientific assessments, politicians propose soft measures that will not disturb the business-as-usual. To be *politically correct*, they must give the impression that they will solve the problem, so they promise large emission reductions in 20 or 30 years.

- These politicians then do nothing to implement the needed structural changes.

- Environmental groups propose actions that are more ambitious than those of politicians. But to simplify their strategy, they focus 95% of their efforts on a few symbolic messages. Even if they would deny this, their overall message becomes *oil and plastic are the villains; wind power and electric vehicles will save us from climate change.*

- A group of citizens reads about climate change; they become worried, feeling guilty about consuming oil or electricity. Another group of citizens is conservative and see climate measures as a personal attack on their lifestyle.

- Promoters of *politically correct* technology then intervene. They propose options that allow citizens to feel less guilty, without changing their lifestyle: one example is electric cars. They also offer options that will enable big companies to improve their image: for example, wind or solar power.
- Environmental groups intervene by also proposing actions that allow citizens to reduce their feeling of guilt: examples are residential recycling or reusable grocery bags.
- Promoters of new technology do not want to offend conservative citizens. They say that managing climate change is easy; they promise that wind power and electric cars will be available at no extra cost.

This pattern has repeated itself at least five times in the last 30 years, without any serious emission reduction. It is surprising that most environmental groups and politicians just want to repeat it again.

Climate change is a collective challenge

When reading this book, we hope that citizens understand that environmental protection is NOT a religion where citizens are guilty or innocent. On numerous occasions, we will insist that we are NOT attributing blame. To win the battle against climate change, we need to change the **collective economic rules.** When climate actions are only proposed to reduce feelings of guilt, their contribution to emission reduction may be insignificant.

It is essential to separate collective and individual issues. A citizen can own a vehicle because it is useful, while still being a promoter of a carbon tax and public transit.

This book is unique because it examines how *political correctness* will probably doom modern civilization. Trying to be optimistic, this book also shows that we can save our civilisation, on one condition: eliminating *political correctness* within climate policy.

The false need to stay optimistic

Despite the solutions presented in this book, some climate change militants will say it is too pessimistic. This fails to recognize the failures of the past 30 years:

- Climate experts have put the emphasis on rational analysis, presenting optimistic scenarios;
- Environmental groups have insisted on popular technological options (windpower and electric cars) that would solve all energy problems, without affecting citizens' lifestyles;
- Countries have not respected their commitments, resulting in non-stop growth of greenhouse gas emissions.

To explain these failures, we can use the lessons of political science. During an election campaign, candidates can use three strategies to convince citizens to vote for them:

- Use a rational approach to show they are the best candidate.

- Appeal to the voter's sense of duty, explaining why they should vote for a candidate who will not waste public money.
- Use emotional arguments, like declaring that the election will determine the well-being of their children or help them avoid the threat of immigration.

Based on political science, a successful political campaign should have the following balance: 10% of the efforts should be on rationality, 30% on citizen's sense of duty, and 60% on emotional arguments.

For the past 30 years, climate change experts and militants have focused on rational arguments and a little bit on the sense of duty. They have avoided talking about issues that might produce an emotional reaction. **This is a major communication mistake**.

In Chapter 1, we will show that unchecked climate change will have huge impacts. We could do like other climate experts who continue to talk of *zero emissions*, in a cold and rational approach. But the reality of climate change is scary. **Now is the time to change strategy, and use emotional arguments. Fear is an emotion that can generate action.**

1

Climate issues greatly underestimated by *political correctness*

1.1 The IPCC, a key institution that underestimates the speed and impacts of climate change

About every six years, the Intergovernmental Panel on Climate Change (IPCC) produces an extensive assessment on climate change. Their reports are the most influential in defining the issues and solutions.

It is essential to understand **how** these reports are produced. For each chapter, numerous authors from different countries are nominated to summarize the state of energy and climate science. Each chapter is then subjected to the following sequence:

- Lead authors write a confidential *first-order Draft* with help from contributing authors;
- The authors then receive comments from independent *Expert Reviewers*;
- A *Second-order Draft* is produced;

- It is reviewed and commented by governmental experts from 180 countries;
- The *Third-order Draft* is then produced and edited to the final version.

At each level, the statements must become acceptable to all reviewers. Unavoidably, this leads to watered-down versions. Risks and impacts are reduced to generalized statements. As I was nominated *Expert Reviewer* for three IPCC reports, it was easy to confirm this watering-down. Here is one example of modification implemented by the process:

- *First order Draft* of the 2005-2007 assessment: Within a high emission scenario to 2060, the report predicts, with a high level of confidence, that most coral reefs will be destroyed.
- Final report (2007)[4]: There is a probability that many reefs will be destroyed.

In 2020, research confirmed that the 2005 *First order draft* was accurate: with a warming of only 1.1 degrees Celsius, it concluded that already 50% of coral reefs had died. A New Scientist article (2021)[5] on biodiversity started with the following sentences:

> *The Great Barrier Reef is already in a critical state. Rising sea temperatures are killing corals faster than they can recover.*

---

4   Climate Change 2007 - The Physical Science Basis Contribution of Working Group I to the Fourth Assessment Report of the IPCC.

5   New Scientist, How climate change hits nature, April 10, 2021, p. 41.

Coral reefs are not an exception. The IPCC has constantly underestimated the impacts of climate change. Another example is the thawing of summer Arctic Sea ice:

- The first two reports (before 1996)[6] made predictions concerning the loss of arctic sea ice: impacts would become significant after the year 2100.
- The third report (2001)[7] included model runs: effects would become significant between 2040 and 2060. The impacts of global warming had moved closer by 50 years.
- The fourth report (2008)[8] compared the model runs with reality. In 2007, the actual ice loss was at the same level as the previous prediction for 2050. Impacts had moved closer by another 40 years.

The IPCC also underestimated the effects of warming on the Antarctic:

- In the first four reports, the IPCC concluded that global warming was not a significant threat to the Antarctic. The third report[9] concluded that *The Antarctic ice sheet is likely to gain mass because of greater precipitation.*

---

[6] Intergovernmental Panel on Climate Change, Climate Change 1995.

[7] IPCC, 2001: Climate Change 2001: Synthesis Report. A Contribution of Working Groups I, II, and III to the Third Assessment Report of the Intergovernmental Panel on Climate Change, Cambridge University Press, Cambridge, United Kingdom, and New York, NY, USA, 398 pp.

[8] Climate Change 2007 - The Physical Science Basis Contribution of Working Group I to the Fourth Assessment Report of the IPCC.

[9] IPCC summary presentation by U. Cubasch, IPCC, 2001.

- In 2021, the IPCC completely reversed its assessment of the Antarctic ice sheet[10]:

  Mass losses from West Antarctic outlet glaciers, mainly induced by ice shelf basal melt (high confidence), outpace mass gain from increased snow accumulation on the continent (very high confidence)…

  there is very high confidence that the Antarctic Ice Sheet lost mass between 1992 and 2017.

1.2 A serious IPCC neglect: climate-feedback mechanisms

Many ecosystems have stored large amounts of carbon over the past centuries, either in the soils or in the forests. In some cases, the warming climate can return some of this carbon to the atmosphere, accelerating global warming. This additional warming then releases more carbon, causing feedback that further increases warming.

Some feedback mechanisms are dangerous because a small warming can release large amounts of these carbon stocks. In all their reports before 2021, the IPCC did not include feedback mechanisms into their global warming predictions. In 2021, the Sixth Assessment Report did discuss many

____

[10] IPCC, 2021: Summary for Policymakers. In: Climate Change 2021: The Physical Science Basis. Contribution of Working Group I to the Sixth Assessment Report of the Intergovernmental Panel on Climate Change, Cambridge University Press, section TS-33 and p. 3-54.

climate feedback issues. But only one of them is included in the warming prediction: cloud feedback which is a minor issue. There are major feedback issues that are still being neglected.

One feedback process is already happening. In 2019, National Geographic[11] explained the warming of the Arctic Ocean:

> *A Vicious Cycle. Snow and ice reflect most incoming light, but open water is less reflective, so it absorbs more heat. More melting causes more open water, a feedback loop that leads to even more warming.*

The Arctic seawater retaining more heat, it will thaw more ice in the next season.

The most worrying feedback process is **Arctic methane release**. Methane is a powerful greenhouse gas; over 20 years, it has a warming potential about 80 times stronger than carbon dioxide (per gram). Arctic permafrost and marine sediments contains huge amounts of methane. Global warming will thaw the permafrost, releasing large quantities of methane. This methane will then accelerate global warming and the thawing of permafrost. The 2019 National Geographic article[12] described this risk:

> *The Thaw Speeds Up. The unexpectedly rapid collapse of ice-rich permafrost in the Arctic could pump billions of additional tons*

---

[11] National Geographic, September 2019.
[12] National Geographic, September 2019.

of methane and carbon dioxide into the atmosphere every year – a threat that has yet to be accounted for in climate models.

In 2021, the IPCC assessed this concern. They recognized that it is a major risk, but it was not included in warming assessments because of insufficient data[13]:

In conclusion, there is high confidence that the permafrost region has acted as a historic carbon sink over centuries to millennia and high confidence that some permafrost regions are currently net sources of CO_2. There is robust evidence that some CH_4 emissions sources and for some regions have increased over the past decades (medium confidence).

They also explain why this process was not included within global predictions:

In addition to CO_2, CH_4 emissions from the northern permafrost region contribute to the global methane budget, but evidence as to whether these emissions have increased from thawing permafrost is mixed.

The shallow subsea emissions are particularly uncertain due to both the wide range of

---

13 IPCC, 2021: Summary for Policymakers. In: Climate Change 2021: The Physical Science Basis. Contribution of Working Group I to the Sixth Assessment Report of the Intergovernmental Panel on Climate Change, Cambridge University Press, p. 5-64 and 5-65.

> *estimates (...) and the lack of a baseline with*
> *which to infer any changes.*

This is consistent with past practices: Whenever there is uncertainty, the IPCC always chooses to minimize the risks of climate change.

Another climate-feedback is related to forests that capture and store significant amounts of carbon. Global warming will cause desertification, forest fires, and the loss of many forests. This will return much carbon to the atmosphere, accelerating the warming trend.

The survival of the Amazon forest is important because it stores large amounts of carbon. The IPCC Sixth Assessment Report, within a high emissions scenario, indicates that there will be an overall drying of the Amazon[14]. Despite this, they do not include this risk within future warming scenarios.

1.3 IPCC Fifth and Sixth Assessments: a high level of *political correctness*

Politicians are often pressured by environmental groups concerning the level of GHG emissions in their countries. In response, politicians make (often empty) promises expressed in the percentage of national emissions.

---

[14] IPCC, 2021: Summary for Policymakers. In: Climate Change 2021: The Physical Science Basis. Contribution of Working Group I to the Sixth Assessment Report of the Intergovernmental Panel on Climate Change, Cambridge University Press, figure SPM.5 Change in annual mean surface temperature, precipitation, and soil moisture.

From a scientific perspective, one major issue is how these political promises will translate into specific global warming. One priority of the IPCC should be to clearly show the impacts of future GHG emissions, especially if they are not greatly reduced. Unfortunately, it seems that the IPCC has decided to avoid this, as it could be bad for the image of many politicians.

It is also interesting to compare the various IPCC reports, to show that *political correctness* is increasing. In the Third and Fourth Assessments, it was easy to make a relation between emission scenarios and the level of warming.

A business-as-usual scenario would lead to a global warming of +5 degrees Celsius (+5°C) around 2100. This high-emission scenario included prudent assessments of large impacts.

> Reminder:
> +1°C = +1.8°F
> +2°C = +3.6°F
> +5°C = +9°F

The Fifth Assessment (2014)[15] decoupled the link between emission scenarios and global warming, creating confusion. It described theoretical concentrations of GHGs (*Representative Concentration Pathways*) that are difficult to relate to emission trends. In official communications, this confusing method was followed by the description of impacts of an optimistic +2°C scenario.

---

[15] IPCC, 2013: Climate Change 2013: The Physical Science Basis. Contribution of Working Group I to the Fifth Assessment Report of the Intergovernmental Panel on Climate Change, Cambridge University Press, Cambridge, United Kingdom and New York, NY, USA, 1535 pp.

The Fifth Assessment discussed large emission reductions at the world level: Stabilizing climate at +2°C involves the need, by 2050, to reduce world emissions by 50% and to achieve zero net emissions by 2100. This can be misleading because a 50% global reduction does not consider the differences between rich and poor countries. The Fourth Assessment (2007)[16] was more realistic as it distinguished between the goals of rich and developing nations. Their recommendations to achieve stabilization at +2°C were:

- Poor nations need to develop and increase their energy consumption. Despite this, by 2050, they must reduce their GHG emissions relative to 1990.
- By 2050, rich nations must reduce their GHG emission **by 85%** (relative to 1990). This is a huge challenge. Doing better than 85% is unrealistic.

In the Fifth Assessment, even if a +4.5°C warming scenario was considered possible, there was minimal discussion on it.

The Sixth IPCC Assessment report (2021) was affected by a different type of *political correctness.* As the report was being finalized, some politicians and countries were still discussing the goal of stabilizing the climate at +1.5°C. As the IPCC did not want to discredit pro-active politicians, their report maintained that such a goal was achievable. Trying to please some countries, the IPCC even introduced a new much lower emission scenario.

But the emission scenarios of the Sixth Assessment show that the +1.5°C goal for 2050 is impossible to achieve: With an intermediate emission scenario, **+1.5°C is exceeded**

---

[16] Climate Change 2007 - The Physical Science Basis Contribution of Working Group I to the Fourth Assessment Report of the IPCC.

around 2031. Even with the very low emission scenario, the global warming level of +1.5°C is exceeded in 2040[17].

1.4 Reality: Even with progress in energy efficiency, energy consumption will grow in many nations

The following trends justify an increase in energy consumption:

- The world's population is growing.
- In developing nations, there are one billion people with no reliable electricity generation (or none at all). They want essential energy services.
- Citizens in developing nations aspire to a Western country lifestyle. This implies growth in energy consumption and GHG emissions. Meat consumption will increase in developing nations, causing significant GHG emissions.
- There are numerous reasons (including climate change) why a large number of citizens will migrate from poor to rich countries. These migrants will increase their energy consumption.

---

[17] IPCC, 2021: Summary for Policymakers. In: Climate Change 2021: The Physical Science Basis. Contribution of Working Group I to the Sixth Assessment Report of the Intergovernmental Panel on Climate Change, Cambridge University Press, TS.1.3, Cross-Section Box TS.1.

1.5 Reality: Many proposed actions neglect life-cycle emissions

> *Life cycle assessment (LCA) is a methodology for assessing environmental impacts associated with all the stages of the life cycle of a commercial product, process, or service[18].*

For example, the life-cycle of natural gas will include extraction, delivery, and combustion of gas. We will not elaborate here, as the next six chapters of this book are dedicated to life-cycle assessments.

But there is one obvious fact: *Zero-emission* cars do not exist, as we must consider the emissions created in producing the car, building roads, and providing the electricity required to fill the batteries.

Moreover, life-cycle assessments show that the concept of *zero-emission* may never be achievable for numerous activities. Here are a few examples:

- Rice is an essential crop that is produced on flooded soils. Any decomposition of organic matter, notably in rice production, produces methane, a powerful greenhouse gas.
- Based on current technology, there is no zero-emission scenario for commercial aircraft.
- Vaclav Smil, an expert in energy transitions, states clearly[19]: *An even greater challenge will be to displace*

---

[18] Wikipedia.
[19] Vaclav Smil, Examining Energy Transitions: A Dozen Insights Based on Performance, 2016.

> *fossil carbon used to produce iron, cement, ammonia and plastics.*

- It is difficult to imagine that countries will handicap their military strength by adopting *zero-emission* war equipment. Military airplanes will continue to fly.

Expectations about *zero-emission* technologies are greatly exaggerated.

1.6 Reality: Governments continue to subsidize polluting activities, despite past commitments

Many reports, notably by the International Energy Agency, have confirmed that the oil and gas industries receive large subsidies. Moreover, private transportation receives subsidies in the form of free roads and free parking. Because pollution is subsidized, it is unrealistic to assume that individual choices can have a major impact.

Many books propose that people change their lifestyles to make large emission reductions. Within the current economic context, this is like asking kayakers to paddle upstream against a strong current. First, we must change the context to allow kayakers to paddle downstream.

Because of this reality, we will propose fiscal or economic reforms, to reward citizens who reduce their GHG emissions. Such reforms should be at the top of political agendas at all political levels.

1.7 Stabilizing the climate at +1.5°C is wishful thinking

At the 2015 United Nations Climate Change Conference in Paris, the main debate was "Should we limit global warming to +1.5°C or +2°C". This debate was a waste of time for many reasons:

- The current commitments of countries cannot limit warming at +2°C.
- Assuming all commitments are respected, the UN Environment Programme concluded that we are on a path to +3.2°C[20]:
- *Even if countries meet commitments made under the 2015 Paris Agreement, the world is heading for a 3.2 degrees Celsius global temperature rise over pre-industrial levels, leading to even wider-ranging and more destructive climate impacts.*
- In the past, most countries did not respect their **modest commitments**. Now that commitments are ambitious, is it reasonable to assume that they will ALL be respected?
- Many recent commitments are unrealistic. They will be questioned by each newly elected government. Incoming right-wing governments may reject them.
- The global warming assessments do not include many climate-feedback mechanisms.
- The IPCC authors do not want to offend the politicians who fund their research or pay their salaries. The goal of stabilizing climate at +2°C is nearly impossible to achieve, but the IPCC will not say it explicitly. And

---

[20] U.N. Emissions report, November 2019.

there is a big difference between the +1.5°C and +2°C goals: To respect a +2°C commitment, we need zero net emissions by about 2100; **to respect +1.5°C, we need zero net emissions by 2050**.

In 2020, about 45% of the world's population lived in China, India, and Indonesia. These countries produced more than 60% of their electricity with coal. It is ridiculous to assume that these developing countries will achieve zero net emissions by 2050. Stabilizing climate at +1.5°C is not realistic.

Each IPCC report was composed of three volumes, the third dedicated to emission reduction measures[21]. In these volumes[22], the IPCC has failed to show the real issue: Without a deep transformation of economic and fiscal systems, we are heading to a +5°C warming.

---

[21] Climate Change 2007 - Mitigation of Climate Change Contribution of Working Group III to the Fourth Assessment Report of the IPCC.

[22] IPCC, 2014: Climate Change 2014: Mitigation of Climate Change. Contribution of Working Group III to the Fifth Assessment Report of the Intergovernmental Panel on Climate Change, Cambridge University Press, United Kingdom and New York, NY, USA.

1.8 Reality: We are on a path to a +5°C scenario, with very large impacts

The IPCC Fourth Assessment (2007)[23] made one serious contribution: showing that global warming of only +0.8°C was already multiplying numerous impacts:

- More intense and longer droughts.
- Increases in extreme temperatures, frequency of heavy precipitation events, and tropical cyclone intensity.

For the future, it showed that business-as-usual leads us to +5°C. It concluded with the following impacts of a +4°C warming (A2 emission scenario):

- Major species extinction at the global level.
- More intense hurricanes and heat waves.
- Heat waves and heavy precipitation events will continue to become more frequent.
- Future tropical cyclones will become more intense.
- It presented a picture of a worrying emission scenario (A2), that leads to of the +4°C warming for the year 2090.

---

[23] Climate Change 2007 - Impacts, Adaptation and Vulnerability Contribution of Working Group II to the Fourth Assessment Report of the IPCC.

IPCC AR4 (2007) warming scenario (°C)[24]

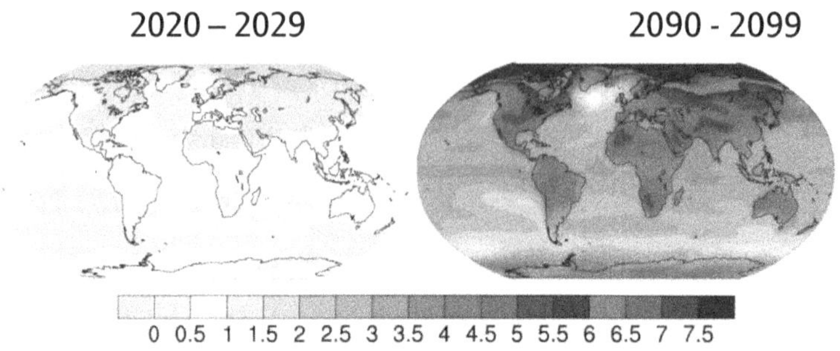

It is essential to repeat that this +4°C scenario does not include climate-feedback mechanisms. With the combination of business-as-usual and feedback mechanisms, a +5°C scenario is probable.

In Assessment Report no. 5, the IPCC did not discuss the implications of such a +5°C warming. Many issues were neglected or downplayed:

- They did not say that heat waves will be so intense in some regions that human habitat will become nearly impossible.
- They did not assess the combined effects of rising sea levels, stronger hurricanes, and more intense precipitations; these effects will flood large coastal areas so often that they become inhabitable.

---

[24] Climate Change 2007 – Summary for Policymakers, Fourth Assessment Report, IPCC, figure SPM6.1

1.9 Social-feedback mechanisms and the collapse of modern civilization

We must also consider social-feedback mechanisms. This is not part of the IPCC mandate.

With climate change, the environmental conditions favorable to agriculture will change globally. To understand this effect, we can examine the case of the civil war in Syria, in the context of global warming of +1°C[25]:

> *Starting in 2006, Syria suffered its **worst drought in 900 years**; it ruined farms, forced as many as 1.5 million rural denizens to crowd into cities alongside Iraqi refugees and decimated the country's livestock. Water became scarce and food expensive. The suffering and social chaos caused by the drought were important drivers of the initial unrest.*

> *Impacts of the civil war: 400 000 people lost their lives. There were 5.6 million refugees, out of a total population of 22 million.*

The Syrian civil war and other Middle East problems created millions of refugees, many wanting to migrate to rich countries (mainly Europe). Some Western governments used this refugee crisis as justification to raise barriers to immigration.

Let's move to 2080, with global warming of **+5°C:**

---

[25] PBS, March 17, 2016, PBS Internet site.

- Heat waves have rendered many regions nearly inhabitable. This includes numerous large cities in India and the Middle East.
- Rising sea levels, stronger hurricanes, and more intense precipitations have caused repeated and massive flooding of coastal cities. The majority of ports, including related industries, need to be moved to higher ground. Huge coastal areas built for tourism are abandoned, notably tropical islands and South Florida[26].
- Droughts have greatly reduced agricultural production. Many regions cannot produce at all due to a lack of rainwater and no more irrigation. There is widespread political tension because some temperate countries (where agriculture was less affected) have decided to limit their exports of cereals.
- Fisheries, once a major source of protein, have collapsed, as many fish stocks has disappeared, due to warming waters and overexploitation.
- Most developing nations have seasonal famine conditions.
- In many cities, drinking water is limited, and water riots are common.
- Civil wars have erupted in most developing countries.
- Many millions of refugees want to migrate to rich countries.
- There are also large migrations within countries. In the U.S., millions of citizens need to move away from coastlines, and a large portion of the West has become nearly inhabitable because of desertification

---

[26] National Geographic, February 2015.

and lack of water. Political divisions become even more extreme and violent.

Here are the social-feedback processes involved:

- Flooding of coastlines, lack of water in dry regions, low agricultural production, and famine cause massive migration. These famine and migration cause tension within or between countries. Political elites react by building up military capacity and increasing repression.
- Political systems become more totalitarian, inciting even more unrest, favoring long-term civil wars.
- Civil wars cause further decline in agricultural production.
- All these impacts cause the economic collapse of numerous industries because of plant closures and fewer customers. The only industry that is thriving is military production.
- As political systems react aggressively, they cause more migrations and more economic collapse.

In Syria, it was easy to recognize social-feedback processes due to water and food shortages.

Jared Diamond has analyzed the collapse of civilizations. He made the following warning[27] in 2005:

> *Today, just as in the past, countries that are environmentally stressed, overpopulated, or both, become at risk of getting politically stressed, and of their governments collapsing. When people are desperate, undernourished,*

---

[27] Jared Diamond, Collapse, How Societies Choose to Fall or Succeed, p.516.

> *and without hope, they blame their governments, which they see as responsible for or unable to solve their problems. They try to emigrate at any cost. They fight each other over land. They kill each other. They start civil wars.*

With global warming of **+5°C**, migrations and refugees will probably be 100 times greater than those from the Middle East during the 2015-2019 period. With social-feedback mechanisms, refugees will be counted in hundreds of millions, maybe billions of citizens. This will cause the collapse of modern civilization.

1.10 The next critical challenge to the IPCC

For each IPCC report, the lead authors produced a *Summary for policymakers*. For six reports over 25 years, the authors adopted a moderate tone; they did not convince policymakers that managing climate change is a priority. To be more convincing, the IPCC needs to change its tone and strategy. **If managing climate change is to become a priority, we must tell the truth to policymakers: We are on a path to at least +4°C, probably +5°C. Such warming will cause the collapse of modern civilization.**

The next six chapters of this book are dedicated to another problem: the green performance of many new technologies is greatly overestimated by *political correctness*. Impacts are underestimated, and the benefits of technology are overestimated. This is a recipe for disaster. ***Political correctness* is dangerous: it has lured us into false perceptions that will ensure failure.**

2

The need for clean electricity, underestimated by a factor of three

2.1 Key Questions, answered by political correctness

Two quotes from the International Renewable Energy Agency[28], a promoter of wind and solar energy:

> *-The Paris Agreement sets a goal to limit the increase in global average temperature to well below 2°C above pre-industrial levels and to attempt to limit the increase to 1.5°C. Implicit in these goals is the need for a transition to a low-carbon energy sector, which accounts for two-thirds of global emissions. Renewable Energy, coupled with energy efficiency gains, can provide 90% of the CO_2 emissions reductions needed by 2050.*

> *-The cost of wind power will continue to drop... In 2015, the levelized cost of wind was 7 /kWh. It will be 5.2 /kWh in 2025.*

---

[28] IRENA internet site, 2021.

A few quotes from Greenpeace:

> *-Hydropower and geothermal, the wrong Renewable Energy: China, Japan, and South Korea's RE project finance has concentrated in environmentally and socially damaging hydropower plants ... These neglect serious issues in cost, public health, sustainable economic development, impact on livelihoods and ecosystems, and domestic energy security.*[29]

> *-Nuclear power poses serious threats to the environment and to humanity, and so has no place in a clean energy future.*[30]

These statements are misleading. We will describe why in the next sections.

---

[29] Greenpeace Japan, "Achieving Net-Zero with China, Japan, and South Korea", 2020 p.6.

[30] Internet site of Greenpeace Canada.

2.2 Misleading statistics on electricity

Three technical concepts are needed to understand the marketing of wind and solar power. A power plant has a **maximum power** that it can deliver instantaneously (expressed in MW). The plant produces a **quantity** of electricity yearly (expressed as MWh/year). The linkage between these two concepts is the capacity factor, which defines the quantity of electricity that will effectively be produced on average.

> **Technical issue: Power versus Electricity**
>
> A power plant has a maximum **power** that it can deliver instantaneously. This is expressed in watts, for example, one megawatt (MW) or one gigawatt (1 GW = 1000 MW). The quantity of **electricity**, produced in a year, is expressed in MWh or GWh. For example, if a one MW plant operates at maximum capacity for 4000 hours during a year, it will produce 4000 MWh of electricity.
>
> For this plant, we can calculate the *capacity factor* based on a theoretical yearly maximum of 365 days multiplied by 24 hours/day. The plant produces 4000 MWh out of a theoretical maximum of 8760 MWh/year. The capacity factor is 46% (4000 /8760 x 100).

Because of low wind or lack of sunshine, the capacity factor is 26% for wind power and 13% for solar power.[31] A typical gas-fired power plant has a capacity factor of 80% or more. To replace the electricity from a 1000 MW gas plant, more than 6000 MW of solar panels are needed. What do promoters do? To give the impression that wind and solar development are important, they present data on maximum installed power (MW), not electricity produced. When showing the contribution of renewables, we should always provide statistics on electricity production, not maximum power.

But even statistics on electricity production are biased in favor of windpower, as a windmill will produce intermittent electricity based on the strength of the wind; there may be no correlation between when electricity is needed and when the windmill does produce electricity. If a region has a lot of wind power, other options need to produce more electricity, when the wind is low. In contrast, a hydro plant

---

[31] IRENA, 2018 data.

with a reservoir can produce electricity exactly when needed. Numerous documents and presentations of IRENA do not discuss this intermittent issue, as though it does not exist.

Another misleading statistic is the rate of growth. Promoters of wind and solar do a lot of marketing with their rate of growth[32]:

> *The solar PV market was up. 25% over 2014 to a record 50 GW, lifting the global total to 227 GW. The annual market in 2015 was nearly 10 times the world's cumulative solar PV capacity of a decade earlier.*

But when you start small, the rate is based on a small number, showing a large relative growth, even with small absolute numbers. Based on IRENA data between 2010 and 2018, solar had a growth of 1500% over eight years. In contrast, the growth of hydro was only 20% but, in absolute value, it was much greater than solar.

2.3 Underestimating the future needs of clean electricity

When discussing the future contribution of renewables, the data generally **focus on current coal and gas-fired generation**. Promoters of wind and solar demonstrate that wind and solar can replace all this electricity in a few decades. This approach allows them to say that new renewables are the key to respecting the Paris Accord commitments.

----

[32] Ren21, Renewables 2016, Global Status report.

This is doubtful: The IRENA site shows that wind and solar have increased by 1448 TWh between 2010 and 2018[33]. To replace fossil electricity generation, we need **11 times** that eight-year effort. If the deadline is 2050, this seems possible but would be difficult. And promoters do not provide a clear answer on how to compensate for such large quantities of intermittent electricity. What is the option for ensuring security of supply at night or when there is no wind? What is the cost of this other option?

Another issue is more obvious: **electricity represents about 20 to 25% of total primary energy demand**. To reduce GHG emissions massively, clean electricity is also needed to replace gas and oil in heating activities, gas in industries, oil in transportation, etc.

In absolute numbers, **the need for clean electricity is 38 times the 2010-2018 growth in wind and solar energy** (technical explanation in next paragraph). In this context, it is clear that wind and solar are insufficient. We need a large contribution from energy efficiency, AND other generation options. The electricity from these other options will need to be more constant than electricity from wind and solar power.

Technical explanation: The efficiency of renewables does help

Expressed in primary energy, replacing all fossil fuels appears to be a challenge about five times greater than just replacing gas and coal generation. We need, however,

---

[33]   IRENA internet site, Installed capacity trends.

to use a life-cycle approach. For oil and gas, we need to consider the energy spent in extraction, transportation, refining, etc. Fortunately, over their life-cycle, renewable options are more efficient than fossil fuels. We can make a rough assessment[34], assuming global efficiencies of 80% for renewables, 60% for gas and oil heating, and 20% for oil in transportation.

When factoring these efficiency gains, to replace all current fossil use, the overall challenge is about 3.5 times the electricity sector.

The following table[35] confirms that wind and solar power have much lower emissions than coal. It also shows that hydropower and nuclear energy can reduce GHG emissions.

Reminder:
1 metric tonne
=
1.1 U.S. ton

---

[34] Assessment based on data from the World Energy Outlook 2018.

[35] L. Gagnon, C. Bélanger, Y. Uchiyama, "Life-cycle assessment of electricity generation options: the status of research in year 2001", Energy Policy, vol.30, no. 14, p. 1267-7 Updated version for courses at École de technologie supérieure, 2015.

Life-Cycle GHG Emissions per Unit of Electricity

	Tonnes CO2 eq. / GWh	
	Average existing technologies and sites	Best available technology and best sites
Coal	1022	941
Diesel	787	649
Natural Gas C.C. turbines	499	422
Coal with Carbon Capture	300	220
Solar Photovoltaic	121	38
Wood Plantation	90	51
Windpower	20	9
Nuclear Energy	16	6
Hydro with reservoir	33	10
Hydro Run-of-river	4	3

N.B. Renewable sources depend more on site quality than on technology

2.4 Windpower: unrealistic promises and expectations

Because of previous comments, some people may conclude that we do not support windpower. This is not the case: windpower can make a large contribution to reducing GHG emissions. But many people **overestimate its performance by a huge margin.** This is notably due to unreasonable cost predictions. This is a quote[36] from a well-known promoter of wind power **in 1981:**

> *Studies in Denmark, England, and the United States indicate that electricity that costs between 3¢ and 7¢ per kilowatt-hour should be produced by the next generation of turbines...*

For 2019, IRENA is advertising an average cost of 7¢ per kWh. So declining cost predictions took **40 years** to be realized. This is the story of windpower marketing.

Promoters of wind power still promise significant cost reductions. This is again unrealistic for numerous reasons:

- First, the main reason why costs were reduced in the past 30 years is the size of turbines, evolving from 250 kW capacity to 2 MW or more. The larger turbines benefit from economies of scale and better wind, because they are higher. But there is a limit in growing the turbines in size and height. The very large off-shore wind mills confirm this cost dilemma: they are, per kWh, much more expensive than land-based windmills, despite better wind conditions.
- Second, in the future, the large number of windpower projects will multiply the need for a specific group of metals[37]: Aluminum, Copper, Manganese, Neodymium, Nickel, and Zinc. Many factors point to an increase in the costs of these

---

[36] Christopher Flavin, "A Renaissance for Wind Power", Environment, October 1981.

[37] World Bank, 2017, The Growing Role of Minerals and Metals for a Low Carbon Future.

metals: larger consumption of each metals create economic scarcity, increasing costs; current mining practices are often dangerous for workers and will need to be improved; the best mining sites are the first to be exploited and future sites will require larger investments per unit of production. All these factors will increase the cost of the metals needed by windpower.

Moreover, the cost of wind must include the hidden costs due to its intermittency. It is important to understand that the intermittent character of wind power is not composed of small gradual fluctuations. The next graph presents a typical generation profile from a 100 MW project located on a good wind site, for the best period (a month in winter). Within a few hours, the generation can change from near zero to near maximum, with a return to zero in the next few hours.

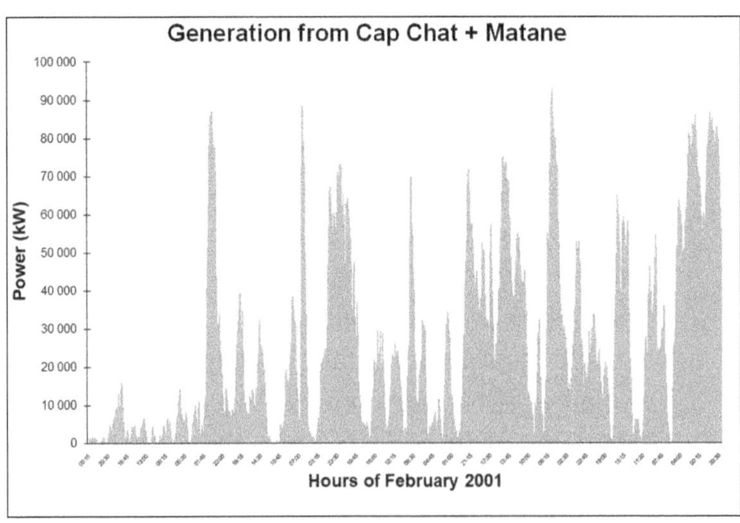

Other equipments are required to ensure the reliability of electricity supply. There are a few possibilities:

- A generation option that can be started quickly when the wind falls. This option could be a hydropower plant with a reserved quantity of water in a reservoir; this is probably the best option, but many regions do not have this possibility. Another option is a gas turbine, with some fuel in reserve. These options have significant costs. Some studies have shown that a gas turbine is often the least-cost option, adding 2¢ or 3¢/kWh to the wind/gas turbine duo. But gas turbines are fuelled by fossil fuels and may not be acceptable within a strategy to reduce GHG emissions.
- Another option to compensate for wind fluctuations could be energy storage. There are numerous options: storing electricity in batteries; producing and storing hydrogen that can later, fuel a gas turbine; pumped storage hydropower. All these options carry a substantial cost, surely higher than 4¢/kWh in the overall cost of the wind/storage duo.

To attenuate the intermittency of wind, some promoters propose a different strategy: build wind capacity, spread out over a large territory. In this scenario, it is less likely that all the wind capacity is affected by low wind at the same time. This appears to be a good strategy, to reduce the size of backup power requirements. In the next paragraph, we explain why this will multiply transmission needs.

Why a major windpower scenario, over a large territory, multiplies the need for transmission lines?

> - The good wind sites may be far from consumption centers.
> - Because of the low capacity factor of wind, transmission lines must be *oversized*. For example, if a gas plant of 20 MW is installed, the transmission line must be capable of carrying 20 MW; in contrast, a wind farm of 100 MW is needed to produce a yearly average of 20 MW. It may produce 100 MW during short periods; so the transmission lines must have higher capacity. If not, a portion of windpower cannot be delivered during windy periods.
> - Transmission line requirements are greater if the wind power is spread out over a large territory. Let's assume there is a strong wind in the Eastern region and none in the West. It means that the lines in the East must be able to carry all the wind energy to all the consumption centers. The same requirement applies in the other direction.

Windpower is needed to replace fossil fuel generation. But the replaced gas and coal-fired plants had one advantage: they were installed close to consumption centers, reducing the length of transmission lines. In comparison, the large wind strategy, spread out over a large territory, will be expensive because it needs **2 or 3 times more transmission lines.**

No matter the costs, replacing coal and gas-fired generation will be essential because of their high GHG emissions. Windpower cannot do that alone, because of its intermittence and hidden costs. In section 2.10, we will discuss another strategy for windpower promoters: instead of *bluffing* on the future costs of windpower, they could promote carbon taxes. This strategy may offend some politicians (not *politically correct*), but it would surely create favorable conditions for more windpower development.

2.5 Solar power: unrealistic expectations

IRENA is predicting a large growth in solar photovoltaic (PV) energy. They also admit to current *levelized* costs of **22¢ to 30¢ /kWh**. This is 3 or 4 times more expensive than fossil fuel electricity. This cost is a big handicap to replacing fossil fuels on a large scale.

Solar PV faces similar constraints as wind power:

- It requires numerous minerals and metals, for which costs are likely to rise.
- Because of its intermittent character, it has hidden costs.
- There is one difference: wind power has the advantage of producing more in winter; it is adapted to cold climate. In contrast, solar PV produces much more in summer and very little in winter. It is adapted to areas that need a lot of air conditioning.

We have seen promoters pushing for solar power in cold countries. In the Canadian province of Ontario, they succeeded, and the government signed contracts at high costs. The provincial government is stuck with these PV projects that will need large subsidies for decades.

The overall issue of intermittent sources has been studied carefully by Vaclav Smil. Analyzing the data of Germany, which has implemented a lot of wind and solar power, he concludes[38]:

So, increased reliance on intermittent energy sources with low average capacity factors for generating power requires countries to maintain fossil-fueled capacity of the same – or even slightly higher – magnitude.

2.6 Biomass energy: reliable, but with smaller GHG reductions than advertised

Electricity from biomass is still a bit more expensive than fossil fuel generation. If fossil fuels are made more expensive by a carbon tax, biomass becomes a serious economic contender.

Relative to climate change, the fundamental debate on biomass energy is related to the regrowth of forest: when you cut a tree and burn it, you release a large quantity of carbon dioxide (CO_2), similar to burning coal; however, if you manage the forest, it will regrow and absorb a similar amount of CO_2. This argument is technically valid but has numerous "if" conditions:

- Will the forest be effectively protected for 50 years, allowing the regrowth?
- Reducing GHG emissions is urgent. Is it acceptable to emit a large quantity of CO_2 now and then wait 50 years to offset it?

------

38 Vaclav Smil quoted in Examining Energy Transitions: A Dozen Insights Based on Performance, 2016.

- What is the energy required to harvest the wood and deliver it to energy consumption sites?

The last question is a key issue in life cycle assessments of biomass projects, as it can greatly affect the performance. Here are examples showing big efficiency differences:

- A pulp and paper mill can use its own waste wood as energy source. There is no additional extraction and transportation of biomass; if the biomass waste is not used, it will decompose and emit GHGs. This case is a true example of zero GHG emissions. Unfortunately, it has limited potential.
- Plantations of fast-growing biomass, with activities similar to agriculture. These plantations require seedlings, maybe fertilizer, weeding, harvesting, and truck transportation to the consumption site. LCAs have looked at various projects[39], with GHG emissions ten times lower than coal but five times higher than windpower. Contrary to marketing, this is far from a zero-emission option.
- Trees extracted from a forest, with mechanized equipment that runs on diesel fuel. Trucks and more diesel fuel are needed to deliver the wood. Freshly cut trees have low energy density, because of a humidity content of 50%. This means that half the truck's load is water, not fuel. If the forest is not allowed to regrow, this option has more GHG emissions than coal-fired generation; even in a well-managed forest, the worst-case scenario can be caused by forest fire. If the forest does regrow, the

---

[39] Matthews, 2000; Styles 2007.

life cycle emissions would be much lower than coal, but again far from zero-emission.

Europe imports large quantities of biomass to replace coal. The GHG emissions associated with producing and transporting the biomass are rarely included in the assessment[40].

Biomass should be part of a strategy to reduce GHG emissions. But, due to these constraints, biomass can only play a limited role. Anyway, forests provide many other environmental benefits that justify their protection.

2.7 Reality: the need to build as much hydro as possible

Life-cycle assessments (LCAs) have shown that hydro has an outstanding performance: reliability of supply, capacity to follow the demand patterns, and longer life than any other option. Typical refurbishment is done after 60 or 70 years; this requires much less money and energy than the original project. Hydropower also has low GHG emissions, similar to those of windpower; it has a very large remaining potential, capable of GHG emission reductions in many countries. But the image of hydropower has suffered because a few projects have caused serious controversies. And local opposition rarely considers the results of LCAs.

Like all renewables, hydropower potential is limited by the quality and quantity of sites. But in contrast with other

---

[40] The great carbon scam, New Scientist, Sept. 24, 2016.

renewables, politicians or environmental groups greatly **underestimate** the hydro potential. It is *politically incorrect* to propose new hydro projects. Alternatively, it is *politically correct* to say that hydro affects local populations or native people. Some promoters of windpower fear the competition from hydro; they say that hydro is not renewable, an absurd argument from a scientific perspective. Considering the great need for renewable energy, there is ample space for both of them.

The case of the U.S. can illustrate how hydro potential is underestimated. In the U.S., the official hydropower potential excludes complete river basins, if a small portion of the river basin touches one of these areas[41]: National Battlefields, National Historic Parks, National Parks, National Parkways, National Monuments, National Reserves, National Wildlife Refuges, National Management Areas, National Wilderness Area, and National Scenic River.

Despite these exclusions, the hydropower potential is important at about **90 000 MW**, which could replace half of all coal-fired power plants in the U.S. But hydropower is *politically **incorrect**,* and most states are not considering its development.

At the world level, large remaining hydropower potential is technically and economically feasible. From an economic perspective, hydropower projects at 10¢/kWh are often considered too expensive, but most wind projects are

---

[41] U.S. Dept. of Energy, "Water Energy Resources of the United States with Emphasis on Low Head /Low Power Resources, April 2004, p.17.

considered "competitive", even when true costs are much higher.

One important advantage of hydropower: water can be stored in a reservoir when the wind is strong, and then the water can flow into hydro turbines when the wind is low. The presence of hydro allows the implementation of more windpower. This is not just a theoretical advantage: many countries that currently produce a lot of windpower, benefit from hydropower to compensate for the intermittence of wind: the province of Quebec produces about 10 TWh/year from wind, thanks to its large hydro capacity; Denmark has a high level of windpower, that is supported by hydro from Norway.

In Canada, four provinces (British Columbia, Manitoba, Quebec, and Newfoundland) have large hydropower capacity providing more than 80% of their electricity. These provinces have postponed hydro development, on the premise it is not needed. This presumption is preventing large GHG reductions. As discussed previously, much energy is not consumed as electricity: more hydro could replace natural gas consumption, with major GHG reductions. Moreover, many neighboring provinces or states still use fossil fuels to produce electricity: more hydro could be exported to reduce emissions.

The main dilemma: climate change will modify nearly all ecosystems on the planet and destroy many; to avoid this, maybe we should accept the effects of renewable energy development on local ecosystems. Historically, many hydro projects have been abandoned because of opposition from local population and environmental groups. And in the past

20 years, many wind power projects have also been canceled because of local opposition. We must accept some local impacts, to win the climate change battle.

The global remaining hydropower potential is large. Many U.S. States and Canadian provinces could exploit their hydro potential to reduce GHG emissions. This is only possible if they set aside *political correctness* on hydropower.

Unfortunately, not all countries have a large hydro potential, notably in dry climates. They will need to produce electricity with another low emission base-load option. In some cases, nuclear energy could be the only alternative to fossil fuels.

2.8 Reality: many countries will need nuclear power

Nuclear energy has minimal GHG emissions and large potential. Many environmental groups and politicians want to avoid this option, at any cost.

To get a rational perspective on nuclear energy, we must compare it with coal, the other base-load option available anywhere. Let's describe the reality of coal, with many impacts neglected by environmental groups:

- Globally, there are 7 million deaths each year due to air pollution[42]. **About 2 million of these deaths, per year, are directly due to coal.**

---

[42] United Nations Environment Program, 2019.

- Per unit of energy, GHG emissions from coal are 2 to 3 times greater than those from natural gas. So a large portion of the impacts of climate change must be attributed to coal.
- When we consider the complete life-cycle, the land-use impacts of coal are many times greater than those of other options: mining of coal (often on large strip-mining areas), transportation of coal by train, acid rain affecting thousands of square miles of forests, lakes, and farmlands; and, in the future, the huge loss of land caused by climate change.

In the last 10 to 20 years, many groups have criticized China because of its nuclear program. But the impacts of coal justified the choice of nuclear. Here are a few statistics (2008) on coal in China[43]:

- Coal was responsible for 70% of soot, 85% of SO_2, 67% of NO_x, and 80% of CO_2 emitted in the country.
- 30% of the land in China is affected by coal-related acid deposition.
- Air pollution reduces agricultural yield by 5 to 10%.
- Mercury emissions from coal combustion were about 303 tons. Mercury causes severe pollution to soil. Each year, 12 million tons of food supplies are polluted by heavy metals such as mercury.
- Between 2000 and 2006, there were 18 516 accidents and 31 064 deaths in Chinese coal mines. This counted for 27% of all industrial accidents and 40% of all industrial deaths over the period.

---

[43]   The True Cost of Coal, 2008.

In China, the air pollution from coal is responsible for about one million deaths per year. Nuclear opponents would respond that China should develop wind and solar power instead. But China actually did. In 2019, wind power electricity production (466 TWh) surpassed nuclear production (366 TWh).

What is the quantity of coal required to produce the same electricity as Chinese nuclear plants? About 350 million tons of coal/year. Assuming that this coal is transported by trains of 50 cars of 60 t/car (3000 t), this would require a permanent delivery of about **14 trains per hour**. Another 7 trains are required each hour to exit the ash from the power plants.

In Europe, Germany has listened to its environmental groups and green party, with the rapid development of wind/solar power and a commitment to retire all its nuclear power plants. The following graph[44] shows the difference between marketing and reality: first, the contribution of wind/solar is still small; second, the growth of wind/solar is similar to the drop in nuclear generation.

---

[44] U.S. Energy Information Administration, from World Energy Balances, 2020, Internet site.

Total energy consumption of Germany

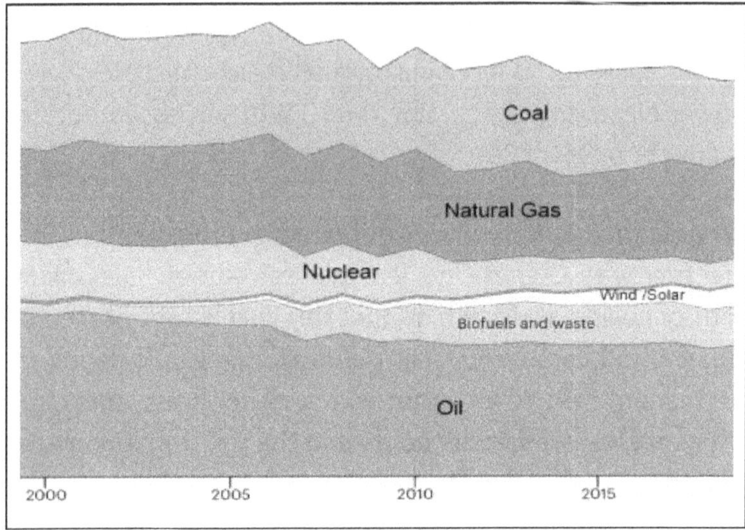

So essentially, wind and solar development has only served to replace nuclear energy. It is also ironic that Germany is buying a lot of nuclear energy from France because of its commitment to close nuclear plants. It is also buying gas from Russia, allowing Russia to build up its armed forces and attack Ukraine.

In 2011, an exceptional tsunami caused the meltdown of one nuclear reactor in Fukushima, Japan. Because of this major accident, numerous groups declared that all nuclear energy should be stopped immediately. After a decade, here are a few relevant statistics on the disaster[45]:

- Reactor-related deaths: one due to radiation exposure.
- Workers exposed to significant radiations: **300**.

---

[45] Wikipedia, 2021.

- Evacuation-related deaths: 2200 due to the power plant meltdown **and** the tsunami.
- The tsunami itself killed about 20 000 people. A few years after the tsunami, some articles created confusion by equating these deaths with the meltdown of the nuclear power plant. This was unfair, as the deaths were caused directly by the tsunami, not the power plant.
- The evacuated area extended 30 km from the plant. This area may appear large, but it is tiny compared to the land-use impacts of coal.

We should compare this exceptional nuclear accident (once every 25 years) with the global deaths caused by air pollution from coal, about 2 million people **per year**. The nuclear debate is not rational.

Nuclear energy does have serious risks. Anti-nuclear advocates cite two major problems:

- *Civil nuclear energy can allow countries to produce nuclear bombs.* This statement is misleading. Natural Uranium is composed of U233, U235 (less than 1%), and U238 (more than 99%). U233 and U235 are the "fissile" atoms that can break and create large quantities of energy. U238 is not fissile; it does not provide energy or cannot be used to make a bomb. Because of this low ratio of fissile atoms, most nuclear reactors consume "enriched" Uranium: this means that a lot of U238 must be removed from natural Uranium, to raise the U235 content to 3% or 4%. When opponents of nuclear energy say that this "enriched" Uranium can make nuclear weapons, it is false. To make a bomb, the U235 content must be raised to more than 90%. This requires a large industrial production dedicated to making bomb-grade Uranium. And bomb-grade Uranium decays over 10 to 20 years, so the production capacity must be permanent. A country that wants to secretly produce a nuclear bomb would need to hide these large industries permanently.

- The other issue that nuclear opponents raise is more serious: nuclear plants produce dangerous waste that must be isolated for thousands of years. The most dangerous and persistent waste is Plutonium (Pu239), which has a high level of toxicity and a very long life of thousands of years. This is a long-term issue that cannot be neglected; unfortunately, because of the past production, a technological solution is needed no matter the future production of nuclear energy. Moreover, **if climate change is not stopped, the modern institutions capable of managing nuclear waste may disappear.**

Closing existing nuclear plants is a big mistake, if coal or gas still serve to produce electricity. In the next decades, some regions will need nuclear energy to reduce their dependence on coal. One example is Australia, where future hydro potential is low, as the flow of many rivers will be diminished by the droughts caused by global warming. Many countries are already mainly deserts. These countries may produce a lot of solar power, but they will need a base load option, to provide electricity during evenings and nights.

2.9 Fusion energy marketing, to avoid the implementation of proven technologies

Many environmental groups insist that nuclear energy is too expensive. In contrast, very few criticisms are directed at fusion research. Promoters of fusion have succeeded in raising large investments. Public investments of $22 Billion currently fund the International Thermo-nuclear Experimental Reactor (ITER).

When it comes to marketing exaggerations, this is even worse than windpower. In the 1960s, there were promises of *unlimited fusion energy in 40 years*. A well-respected textbook[46], published **in 1974,** makes a summary of the mainstream expectations on fusion energy:

> *Whatever approach finally proves effective... it seems likely that a sustained reaction might be possible before 1980, with commercial power available ... between 1990 and 2000.*

Now 50 years later, a sustained fusion reaction is still out of reach. Promoters still promise *unlimited fusion energy in 40 years*. Even if such promises are fulfilled (unlikely), this will be too late to avoid the climate catastrophe.

---

[46] R.H. Wagner, Environment and Man, 1974, p.217.

2.10 France and the analysis of the *Haut-commissariat*

Recently, the planning unit of the government of France[47] has confirmed the difficulties described in this chapter. Here are a few highlights from their assessment:

- To achieve large GHG reductions in the next 15 to 30 years, France must increase electricity generation by 40% to 45%. Such an increase is only possible if new nuclear plants are built to replace those that close.
- To implement an extensive windpower scenario, most wind farms will have to be off-shore. A scenario of 60 000 MW has been proposed. However, the first 500 MW project is facing strong local opposition.
- A massive windpower scenario, within the current network, would create severe problems of electrical stability and security of supply. In the short term, the only option to maintain stability is the implementation of many gas turbines; but this option has a high GHG emission factor, so it is unacceptable. Another option to improve stability would be a very large development of electricity transport and distribution networks, including links with neighboring countries; this is impossible to implement quickly.
- The prospects for solar PV in France are not positive. Any major development would require solar equipment manufactured in China. For many years, this manufacturing would be supported by

---

[47] Haut-commissariat au plan, gouvernement de la France, Électricité: le devoir de lucidité, Ouverture, no.4, 23 mars, 2021.

coal-fired power plants. With the expected energy production from the solar panel (over 25 years), GHG emissions from manufacturing could be greater than the avoided emissions due to the PV installations in France.

- To respect its GHG commitments, France needs BOTH nuclear and windpower developments.

2.11 Serious GHG reductions only possible with a large carbon tax

The true cost of wind and solar energy is still high and will remain high, if we consider the intermittent issue. A carbon tax is essential to create a favorable economic context to replace fossil fuels. Despite the different costs of coal or gas, we must conclude that the carbon tax needs to be large.

We can take Canada as an example: the federal government is implementing a carbon tax of 50$/tonne CO_2 (2022) and is planning to increase it to 170$ by 2030. In Canada, the cost of gas-fired generation was about 6¢/kWh, in 2020. Let's assume that the true cost of wind power (with backup) is 12¢/kWh. To raise the cost of gas at a similar level, a carbon tax of 135$ per tonne of CO_2 is needed. So **the planned increase in the carbon tax should not wait for 2030. It is required now.**

Such a carbon tax can also help justify many hydro projects. The Canadian government does not say that, because of *political correctness.*

2.12 One important political message

From a political perspective, policymakers should understand the following lesson: promoters always paint a rosy picture of their technology, including unrealistic predictions. It is difficult to resist such marketing because entire countries behave like promoters. For example, Germany wants to justify its large investments in solar PV power, so they exaggerate its potential. Twenty years ago, Germany was even a leader in trying to exclude hydropower from the "renewable" list because it is much less expensive than solar energy.

Maybe the lesson can be summarized by a quote from Vaclav Smil[48]:

> *I have never been wrong on these major energy and environmental issues,* ***because I have nothing to sell***.

2.13 Doomed, unless we do this

To prevent catastrophic climate change, energy efficiency programs must make a contribution but will not deliver the needed reductions. We need to urgently replace all coal and gas-fired generation, and many other uses of fossil fuels. The following strategy can be implemented everywhere:

---

[48] American Association for the Advancement of Science, Internet site, Meet Vaclav Smil, the man who has quietly shaped how the world thinks about energy.

1. All countries with hydro potential must implement mega-development programs of hydropower.
2. Huge windpower developments are also needed, including extensive offshore projects.
3. Over the next 30 years, these two options must be capable of replacing all current electricity generation and other combustion of fossil fuels. This is probably possible for Canada, Russia, and Scandinavian countries. But for many other countries, this is impossible; these countries (or states) must consider a mega-development of nuclear energy.
4. This development must be supported by a carbon tax.

Environmental groups and the scientific community should stop judging our climate progress by the empty political commitments on future emissions (in 20 or 30 years). They should judge policymakers by their short-term actions to develop these energy options.

3

Suburban sprawl, the energy consumption multiplier

The previous chapter discussed energy production. The next chapters are dedicated to energy **consumption**. First, we discuss the issue of urban sprawl or, to be more precise, low-density suburban sprawl. Why? Because it greatly affects energy consumption in numerous sectors, notably for transportation and space heating/air conditioning.

3.1 Key urban questions, answered by *political correctness*

What type of building can be considered green? If we look at reports from environmental groups or architects, here are a few characteristics of green buildings:

- Single-family homes with solar panels on the roof.
- Space between houses to ensure maximum solar access or to install windmills.
- Rural surroundings, with many trees.
- Once all this is illustrated, the text is entitled *Zero-emission homes*.

Many things in this picture are false or misleading. In real life, windmills are ten times higher than houses, and each house includes 2, 3, or 4 cars.

Another frequent statement: *Many people prefer single-family houses on a large lot*. We recognize this is true; in this chapter, we will discuss why. The correct statement should be: *Many people prefer single-family houses on a large lot,* **because they don't have to pay for the associated social and environmental costs.** Adding the second part of the statement is *politically incorrect*.

3.2 Misleading perceptions about urban density

The low-density image of a green habitat forgets three issues:

- The public infrastructures required for each house.
- The dependence on numerous private cars and the energy consumption involved.
- It also forgets that the world population is more than 8 Billion: if each household on the planet consumes a large land, there would be nothing left for agriculture or natural ecosystems.

Dense cities are often rejected because of noise and pollution due to traffic. But, very often, the traffic problems are due to the commuting from the low-density suburbs. And when citizens and environmentalists recognize the need to reduce the dependence on private cars, they blame politicians for not providing better public transit. They do not see one obvious consequence of low-density housing: it is impossible to justify a good quality transit system because potential transit customers are too few and far apart. When cities do provide suburban transit with diesel buses, they consume, per passenger, as much energy as a private car (see chapter on transportation for a detailed assessment).

In reality, considering climate change, the total energy consumption should be the first criteria to define if a neighborhood is green.

3.3 The facts about suburban sprawl and urban density

The "car-bungalow-suburb" trilogy multiplies energy consumption for numerous economic activities. Because of political debates on urban sprawl in the 1990s, many publications date from that period. One of my articles is the result of a research project[49] which modeled the length of infrastructure, based on different urban density.

The three following tables list Energy Multipliers, which compare the impact of three land-use options: rows of inner-city condos (three stories high), inner-city single-family homes, and suburban single-family homes. The condo option is the base of analysis (factor of 1).

---

[49] L. Gagnon, The car-bungalow–suburb trilogy, Ecodecision, December 1991.

The Energy Demand Iceberg (part 1)	Energy multipliers		
	City, Dense	City, Single-family	Suburb Single-family
Direct energy consumption increase, as seen by the individual citizen			
• Increased heating of housing units, because there are no shared walls which allow substantial savings	1	1.3	1.3
• Increased use of private cars, because of longer distances to travel and increased number of motorized trips	1	4	20
• Increased rate of ownership of private cars because of greater reliance on private transportation	1	1.5	2

The Energy Demand Iceberg (part 2)	Energy multipliers		
	City, Dense	City, Single-family	Suburb Single-family
Energy in public services			
• Longer roads and highways, more bridges for the suburbs: construction, maintenance, lights, snow removal	1	1	10
• Longer residential streets and sidewalks per housing unit: construction, maintenance, snow removal	1	4	4
• More streetlights per housing unit: construction, maintenance, and electricity	1	4	4
• Longer sewers and aqueducts per housing unit	1	4	4
• More energy in garbage removal	1	4	4
• More motorized delivery services (ex. mail) because of longer networks	1	4	10

The Energy Demand Iceberg (part 3)	Energy multipliers		
	City, Dense	City, Single-family	Suburb Single-family
Energy for utility infrastructure			
• Longer networks of power lines: construction and maintenance	1	4	10
• Longer networks of natural gas: construction and maintenance	1	4	10
Energy in public transit			
• More energy per transit passenger because of fewer users, traffic congestion caused by cars, and increased length of transit networks	1	4	10
• More energy for school buses: longer networks and reduced ratio of trips made by foot	1	6	20

An energy multiplier of 10 is a multiplication of energy demand by 10. Note that the dense urban scenario is not so dense, as it is based on three story buildings.

For all the activities described in the previous table, a rigorous assessment must include the life-cycle assessment of the transportation system:

- Oil exploration and extraction.
- Transportation of crude oil (tankers, etc.).
- Energy spent in refineries and fuel distribution.
- Raw material production (steel, chrome, aluminum, lead, asbestos, glass, plastics, rubber, and textile) for construction of cars, tires, oil infrastructures (pipelines, rigs, tankers).

- Transportation of raw materials.
- Energy spent in assembly plants for cars: industrial process, assembly, and chemical products.
- Energy to build and maintain parking lots.
- Asphalt for longer roads or highways and more parking.

We also need to add the life-cycle energy consumption caused by urban sprawl on utilities:

- Energy to produce metals (copper, steel) for longer power and cable lines.
- Cement and steel for more bridges, longer sidewalks, sewers, and aqueducts.

Loss of good farmland, creating a scarcity.

- Many farms are pushed to less productive soils. Farmers compensate by increased use of fertilizers.

A few courageous researchers have addressed the issue of sprawl. In 1990, Newman and Kenworthy[50] compared numerous world cities; they showed that per capita fuel consumption (for transportation) was **five times greater** for suburban cities (Phoenix, Houston, Denver, Los Angeles, and Brisbane) than for dense cities (London, Paris, Munich, Stockholm, and Tokyo). That study is outdated, but the contrast between these cities would be even greater now, as low-density suburban sprawl has continued in the last 30 years. In 2012, a book by David Owen[51] showed that technological innovation rarely reduce energy consumption, but that dense cities allow for a major reduction.

---

[50]   P. Newman, J. Kenworthy, Cities and Automobile Dependence, 1990.
[51]   David Owen, The Conundrum, 2012.

3.5 The "freedom" to buy a single-family house in the suburbs

In a democratic system, many people will say it is their right to decide where they will live. With *political correctness*, there is no need to add *no matter the social and environmental consequences*.

First, let's look at subsidies to sprawl. The following table is based on research[52] made at Montreal University:

Subsidies to private vehicles in the province of Quebec (not taking into account private costs)	Canadian $ per vehicle /year (2015)
Roads: construction, maintenance	1600
Free off-road parking	1000
Police, firefighters, accident costs	360
Air Pollutants: • Health effects : 100$ • Greenhouse gases (80$ /t CO_2): 250$	350
Environment: resources, water pollution	300
Public + social costs	3610
Amount paid in fuel taxes, registration and public insurance	500
Road congestion costs (Montreal area 2018)	+3000

We can conclude that, in the province of Quebec in 2015, without considering the cost of road congestion, the net subsidy to the average car was about 3100$CDN /year,

---

[52] L. Gagnon, P.O. Pineau, Les coûts réels de l'automobile, un enjeu mal perçu par les consommateurs et les institutions, 2013, HEC-Montréal, Study updated in 2018.

for 15 000 km/year (9 300 miles/year). Relative to other Canadian provinces or U.S. States, Quebec drivers pay more fuel taxes, and drive less. So the net subsidies are surely greater in the rest of North America.

Todd Litman[53] has confirmed this assessment. Based on typical mileage in the U.S., he estimated the following subsidies and social costs per car, adjusted to 2014 U.S. dollars:

Subsidies to private vehicles in the United States (not taking into account private costs)	U.S. $ per vehicle /year (2014)
Roads and land value	$631
External cost of parking	$876
External costs of accidents	$578
Traffic services	$161
Air pollution and GHGs	$768
Noise	$137
Resources externalities	$442
Water pollution and waste	$151
Public + social costs	$3744

These total social costs for Canada and the U.S., do not include the high social costs of congestion.

Moreover, focusing on the *average* subsidy is an understatement for suburban development. Infrastructure requirements depend on the number of cars and the distance traveled. Compared to an urban household, a suburban household living 30 miles (48 km) away from

---

[53] T. Litman, Victoria Transport Policy Institute, *Transportation Cost and Benefit Analysis*, 2009.

employment centers usually has more cars (2 or 3 or more) and may drive four times the average. For such a household, free transportation services and social costs can represent a subsidy of more than **$15 000/year** (in 2020 dollars).

The U.S. is probably the world champion of suburban sprawl. There are three reasons:

- First, the constant development and abundance of free roads and free parking, like in most other Western countries.
- A second incentive to sprawl is unique: Income tax rules allow taxpayers to deduct interests paid on a mortgage of up to $750 000. Up to 100% of the interest paid on a mortgage is deductible from the gross income.
- The third incentive to sprawl, the government allows taxpayers to deduct municipal taxes from their gross income.

For households, what is the yearly value of these U.S. tax exemptions? Let's make simple assumptions: a house of 750 000$; mortgage of 600 000$ with an interest rate of 3% = 18 000$/year of interests; municipal taxes of 10 000$/year. A household with revenues of 350 000$/year pays 32% marginal income tax. The tax exemptions represent a subsidy of 9 000$/year for a household that does not need any subsidy. If they live in a suburb and have two cars, we need to add 15 000$/year for transportation costs paid by various government levels. **With a total subsidy of about 24 000$ per year, it is easy to understand why many American households buy oversized houses in the suburbs.**

The mortgage interest deduction also means that homeowners who have paid their mortgage, after many years, can refinance their current home, and use that money to buy a secondary home for their vacations.

The 24 000$ /year per household is not an exaggeration, as it includes only transportation subsidies and social costs. It does not include the many costs of sprawl, such as longer aqueducts, sewers, electrical, and gas distribution lines.

Many environmental groups are critical of urban sprawl and its impact on car dependence and loss of agricultural land. But, because of *political correctness*, they never dare to question these large, unfair, and inefficient subsidies.

To be clear, blame can be addressed to economists and environmental groups but not to individual households. For them, **with all this government support, it is a rational choice to buy a large house in the suburbs** and eventually a secondary home.

In a democratic system, where freedom is essential, people would say it is their choice. Yes, it is a choice to receive large subsidies; but it is not the choice of citizens of dense neighborhoods to pay large taxes, that become subsidies to the suburbs. Another injustice to consider: people within cities must suffer the pollution from the numerous cars commuting to and from the suburbs. This pollution has serious health impacts.

In sum, suburban sprawl is not an issue that can be corrected by looking at individual choices. It is necessary to revise the tax system, to improve fairness and energy efficiency.

3.6 Infrastructure choices that create dense cities

To improve energy efficiency, subsidies to road building can be transferred to public transit infrastructures within cities. Such investments attract developments, both residential and commercial. This creates a feedback process to improve the quality of neighborhoods. It is a tool to reduce urban sprawl without having to tax suburbs.

To be efficient, such a strategy requires a few minimal conditions:

- New roads, bridges or tunnels should NOT be built in favor of suburbs.
- Public transit investments must not be squandered on a few very expensive options. This means choosing infrastructures adapted to expected transit demand: underground metro only in very dense areas, trams (light rail transit) in medium-density areas, and buses for low-density areas. We will discuss these options in detail in chapter 5 on Public transit.

3.7 Reducing sprawl with municipal taxes and actions

In many countries and states, municipal taxes are based on the value of a property. This type of taxation sends wrong signals in many ways:

- It is an incentive to buy a large lot and install a relatively small house, minimizing the overall value and taxes. Moreover, there is no incentive to develop

available lots between houses, even if expensive municipal services have been installed, for example aqueducts and sewers.
- Renovations will increase the value of a property. Owners may prefer to avoid improvements to avoid a tax increase. The tax system rewards neglect.
- In many downtowns, speculators buy land, demolish old buildings and create a parking lot. This allows them to wait for a future resale with a high profit. They can survive with this strategy, even long-term: without a building on a lot, the taxes are low, and the parking revenues easily cover them. Well-located land can be underused for decades because of such speculation.
- House value taxation is not proportional to the cost of services, such as road maintenance, streetlights, garbage collection, etc.

It is easy to show that taxation based on house value is very unfair. The following table compares two actual developments in Canada.

A high-density project, in Laval Canada, along a train track	**Single-family homes** nearby
• Land 500 feet by 500 feet, on one side of street (150m x 150m) • Two 15 story buildings and four 6 story buildings • 8 condo units per story • Total of 430 units • Units of one or 2 bedrooms • Municipal taxes, excluding school taxes between 2000$ and 2500$ /unit • Total tax/year: almost 1$ Million	• Lots 50 feet wide (15 m) • Houses of 3 bedrooms or more • Municipal taxes, excluding school tax: about 4500$ per house One side of 500 feet = 10 houses • Total tax/year: about 45 000$ For the same linear infrastructures: 22 times fewer taxes are paid

Per unit, the low-density neighborhood needs 43 times more public roads than the high-density development. And even if the owners in the low-density model pay more per unit, as a group, they pay 22 times less municipal tax for similar infrastructures.

Some of the services provided are directly proportional to the road length: road and sidewalk maintenance, snow removal, streetlights. The city must provide 22 times more services, per dollar of revenue, for the low-density area. Other linear services, such as sewers and aqueducts, need to be bigger to support the dense development; despite this, we can safely say that, **per dollar of revenue, the city provides ten times more services to the low-density.** This is unfair to condo owners. Historically, left-wing groups have not looked at this issue because they defend families who are tenants (not condo owners). This is a mistake, as landlords of high-density habitats normally transfer these taxes to the tenants through their rent.

In contrast to North America, many European cities have developed vibrant, dense downtown areas, because they are active in maintaining old properties and redeveloping underused areas. Most of them use a better system of taxation. A few examples from dense cities:

- In Barcelona, owners pay a municipal tax based on their overall wealth.
- Amsterdam was developed while the municipal tax was based on the width of each land, along streets and roads. This resulted in narrow lots with narrow houses, often side by side.

Another option would be efficient and better adapted to North American cities: a significant portion of **municipal taxation based on the size of each lot, no matter if the lot is used intensively or not at all.** Cities could implement gradually this type of tax, replacing a portion of the current system. Speculators would pay more tax, an incentive to build something on vacant lots.

3.8 Reducing sprawl with intelligent invoices from Utilities

In the previous section, we compared a high-density project with single-family houses. That comparison can be applied to Utilities such as electricity or natural gas distribution. To provide electricity to the 430 units in the high-density project, the length of Utility lines would be about 1000 feet, with six major electric connections. When electricity is provided to 430 single-family homes on both sides of a street, it requires 11 000 feet of lines, with 430 electric connections. Considering the lines and connections, the low-density option is probably 15 times more costly than the high-density option.

So Utilities provide large subsidies to low-density suburban sprawl. This applies to electricity, natural gas, and cable.

This unfair invoicing is due to a simple practice: often, only the consumption of electricity or gas is invoiced. Infrastructures are ignored in the invoicing structure. To be fair, they should be part of any intelligent invoicing system.

In fact, the injustice is greater than the calculated ratio, because in high-density developments, most of the electricity distribution is done inside the buildings and is paid by the owners. In a low-density development, most of the distribution is outside the buildings and paid by the utility company.

3.9 Doomed unless we stop suburban sprawl

Local zoning efforts are clearly insufficient to stop low-density suburban sprawl. Fundamental changes are needed, to send a signal that roads and parking have a high cost, economically and environmentally. Chapter 10 on ecological fiscal reform will discuss the relative efficiency of different options.

As the vast majority of roads and parking are free, they represent a large subsidy to the use of cars and trucks, which are used more extensively in the suburbs. In economic theory, there is always overconsumption of free products and services. This is why road congestion is universal.

There are numerous options to prevent the overuse of public roads, such as taxes on fuel or distance traveled. Other options are tolls or parking charges, with automatic billing through a transponder on each vehicle.

Some people prefer single-family homes on a large lot. But if they had to pay for the real costs, they would probably choose a different habitat with less land. Another issue is fairness towards the households that live in dense habitat: probably more than 25% of their municipal and income taxes serve to subsidize low-density developments.

4

Harmful trends in heating and air conditioning

4.1 Key Questions, answered by *political correctness*

When the energy performance of buildings is discussed, we often see the following statements:

- Efficient refrigerators or LED lights can seriously reduce greenhouse gas (GHG) emissions.
- Wood heating is a good heating option because wood is renewable.
- Natural gas should be part of the energy transition to a cleaner future.
- The cost of gas heating is lower than electric heating.
- Electric heating is not efficient.

These statements are false or misleading, because of marketing or *political correctness*.

4.2 Efficient appliances and lights, against the *rebound effect*

A frequent misconception: new technology is so efficient that it greatly reduces consumption and emissions. We discuss two relevant examples.

Environmental groups have made campaigns on efficient LED lights. They allow some GHGs reductions, but are they significant? Relative to the overall energy consumption of buildings, lighting represents about 5% of the total and heating/air conditioning about 70%. As we will see in the next sections, the size of houses has nearly doubled. So efficient lighting can reduce energy consumption by 2-3%.

In contrast, doubling the size of a house can increase its consumption by 140%. Moreover, because LED are so efficient, many households have multiplied their lighting demand, for example, during the Christmas period. So essentially, LED lights allow for very small reductions in energy consumption.

Heat pumps are another example of greatly improved technology.

What is a heat pump?

- Common names for heat pumps are *refrigerators* or *air conditioning* units. When cooling is required, the pump takes heat in one place and transfers it to another place. Air conditioning does not create cool air; it transfers interior heat to the outside. A fridge does not create cold; it transfers the heat inside the fridge to the kitchen.
- Such equipment can be designed to work both ways. If it is very efficient, it can transfer the tiny portion of heat within outside cold air to the interior of a building. These are called air-to-air heat pumps.

A book by David Owen[54] looked carefully at the efficiency gains in residential and commercial refrigeration. In the last few decades, the efficiency of heat pumps for refrigeration has improved by a factor of four. But due to the lower energy costs, overall demand has been multiplied by a bigger factor.

These technology improvements are subject to the *rebound effect*, which can include wasteful practices: in groceries, 30 years ago, refrigerators had doors; now, they are open, increasing the overall energy demand.

Within each household, an efficient refrigerator should reduce energy consumption. But with a constant cost of electricity[55]:

> *the old refrigerator didn't go out of service; it moved into our basement, where it remained plugged in for another 25 years.*

---

[54] David Owen, The Conundrum, 2012.
[55] David Owen, The Conundrum, 2012, p.122.

Here are a few components of the *rebound effect*:

- Because of the efficiency improvements and low electricity costs, consumers increase the consumption of relevant services (more lighting and cooling).
- Another effect is indirect: when savings on energy consumption are effectively achieved, it means that more money is left to buy other goods that may be energy-intensive.

In reality, numerous examples show that efficient technology is not sufficient. An article in New Scientist[56] made a summary of research done on the *rebound effect*. Based on 33 studies on new technology, it concludes that overall, 63% of energy savings were wiped out by the *rebound effect*. This does not mean that more energy efficiency is not required; it simply means that other tools must accompany the new technology, for example, a carbon tax or other fiscal incentive.

4.3 Wood can be renewable but is it acceptable for heating?

Numerous publications favor the use of wood to heat houses in a cold climate. As discussed previously, there are limitations to burning wood:

- Is the forests managed on a renewable basis?
- What is the energy required to harvest the wood and deliver it to homes?

---

[56] New Scientist, Climate targets at risk from efficiency paradox, March 6, 2021.

- One critical limitation is air pollution: wood or biomass combustion emits substantial quantities of dangerous pollutants: particulate matter and volatile organic compounds. A few decades ago, in Montreal Canada, wood heating was the worst source of winter smog, even if only about 5% of houses used it. Recently, rules have been implemented to reduce or ban the use of biomass in cities.

In general, biomass heating is acceptable in rural areas with low housing density: in this case, the pollution concentration will remain low, and biomass can be delivered with short transportation. But in this context, the overall contribution of biomass will be minimal.

4.4 *Political correctness* never considers the size of buildings

Architects are proud of building efficient single-family homes. Many assessments are concerned with house performance, often per square foot (or meter) of floor. There are serious flaws with this approach:

- In the past 40 years, the efficiency per square foot/ meter has improved, but households are buying much bigger houses. In the U.S., the average size of new homes was 1 660 square feet in 1973 and 2 687 square feet in 2015[57]. This is an increase of 62%.
- The number of people per household is also diminishing. In the province of Quebec, between

---

[57] Wikipedia, 2021.

1970 and 2000, the number of people per household went from 4.2 to 2.4. In the U.S, the number of people[58] went from 3.33 in 1960, to 2.53 in 2020[59]. In terms of square feet *per person*, the size of houses has more than doubled, increasing energy consumption per capita.

- Moreover, the definition of household does not consider the increase of secondary homes. In Canada, there were 464 000 vacation homes in 1977 and 823 000 in 1999. In the U.S, the numbers went from 5.5 million in 1990 to 7.4 million in 2016. These numbers do not include vacation homes bought in other countries. Vacation cottages or condos can consume a lot of energy, even if not occupied full time. They justify numerous motorized trips over long distances.
- The definition of a green home is also questionable. They are often located 30 miles (50 km) from city centers or major activities. They maintain the dependence on 2 or 3 cars per household. The emissions from these cars can be many times greater than the emission reduction from the more efficient house. A study done for the State of New York confirms this:

---

[58] Statista Internet site.
[59] Wikipedia, 2021.

Average Urban House Outperforms Even the Greenest Sprawl House with Hybrid Cars[60]

	Energy Use: Million BTU per year		
	Household	Transportation	Total
Suburban Single Family	114	125	239
Green Suburban Single Family	71	87	158
Urban Single Family	101	35	136
Urban Multi-Family	56	26	82

- An apartment in a multi-story building may never be identified as green, even if it consumes much less energy than a single-family "green" home. Adjacent walls and ceilings are more efficient than any insulation. Fewer cars are generally required.
- The study for the State of New York did not consider the length of public infrastructures, such as roads, water, sewers, and public lights. In urban conditions, these infrastructures can be reduced by a huge factor.

As mentioned in the chapter on Suburban sprawl, there is one explanation for this demand for oversized houses: municipal and income tax systems that stimulate such demand.

---

[60] Metropolitan Transportation Authority, State of New York, *Greening Mass Transit & Metro Regions: the Final Report of the Blue Ribbon Commission on Sustainability and the MTA*, 2010, p.35.

4.5 Should natural gas contribute to the *energy transition?*

Because natural gas is cleaner than coal, promoters insist that it should be part of a transition to a low-carbon economy. This made sense 30 years ago when the goals to reduce GHG emissions were in the range of 10% or 20%. It made sense that the remaining 80% of GHG emissions came from the best fossil fuels. But in the previous chapters, we have shown that rich countries must rapidly reduce their GHG emissions by 85%. To achieve this, there is **no place for natural gas combustion**.

The promoters of natural gas are 30 years too late. This is even more obvious when we look at the diversity in the current use of natural gas, notably feedstock for the petrochemical and fertilizer industries. In 20 or 30 years, if we maintain some natural gas extraction, it should be dedicated to these tasks. Just keeping this portion of gas production would emit large GHG emissions. There is no future with natural gas combustion.

4.6 The misleading low-cost of natural gas

Promoters of natural gas often say their option is the least-cost heating option. They forget to mention a few public practices that represent direct or indirect subsidies to gas:

- In most Western countries, gas extraction and distribution networks have been heavily subsidized by governments. Why? Because large industries want to have access to gas, and governments like to support their local industries.

- Recently, most gas extraction is based on the fracking technology that creates numerous local impacts. In many countries, fracking can be done without paying any compensation to the affected local populations.

- Many old gas wells are abandoned and may continue to leak methane. Alberta has approximately 95 000 inactive wells and 69 000 abandoned oil and gas wells. In 2020, the Canadian government implemented a $1.7 billion program to clean up orphaned and abandoned wells. These are public funds.

- Burning gas emits large quantities of greenhouse gases. Natural gas extraction and distribution are also responsible for large methane emissions, a powerful greenhouse gas. Globally, only a tiny fraction of these emissions are subjected to a carbon tax.

4.7 Facts on the performance of heating options

There are debates concerning the efficiency of various heating options. Here are a few facts about the equipment inside each building:

- Gas furnaces, including their chimney, are efficient at 70 or 80%. This should not be confused with the efficiency of combustion, which is announced at 99%.

- Contrary to many statements, electric baseboard heating is efficient at nearly 100%, meaning that all the electricity is converted to heat.
- If the outside air is not colder than minus 10°C, modern electric heating can be done with air-to-air heat pumps that transfer some outside heat to the inside of the building. This option can have an **average** heating efficiency of 200% (or more), as heat transfer can be twice the electricity consumption.
- But air-to-air heat pumps have a weakness: if the outside air is colder than minus 10°C, efficiency goes down to less than 100%. At that level, a conventional furnace or baseboard units must take over. As a result, air-to-air heat pumps will significantly reduce average electricity consumption but not peak demand when it is very cold. This is an issue if the maximum generation capacity can be exceeded in very cold weather.
- If the outside air is often colder than -10°C, more expensive geothermal electric heat pumps are justified. Because they get heat from the ground and groundwater, they remain very efficient in cold weather, and their average efficiency can exceed 300%. If overall peak electricity generation is an issue, geothermal heat pumps are preferable to air-to-air heat pumps.

In reality, calculating the efficiency of options must include their life-cycle performance, which includes fuel extraction, delivery, and electricity generation. If promoters of gas are often able to say that their option is efficient, it is because they compare themselves with coal-fired generation. The next graph[61] does show that the combination of electricity from coal and electric baseboard units has high GHG emissions. This poor performance is due to coal.

Life-Cycle Greenhouse Gas Emissions from Heating Options

Heating Technology	Energy Source	g CO2 eq./MJ
Electric heating from various sources of electricity		
Electric baseboard	Coal	283
	Natural gas turbines	120
	Hydro	6
Air to air heat pump	Natural gas turbines	71
Geothermal heat pump	Natural gas turbines	40
Direct fuel combustion		
Furnace	Natural gas	83
	Oil	112
Wood stove	Wood	16
Wood fireplace	Wood	86

Electric heating performs better when a significant portion of electricity comes from hydro, wind, or nuclear. Even if the electricity comes from natural gas (combined cycle plants), electric heat pumps have a better performance than direct gas heating.

There is no place for gas heating within a future clean electricity context. Moreover, electricity is always the best

---

[61] Fact sheet 2007 on Hydro-Quebec Internet site: Atmospheric Emissions of Heating Options; Updated for Course at École de technologie supérieure, 2015

option for cooling. Globally, gas network exare a mistake, as they will have to be closed in the next 10 to 20 years.

4.8 Carbon taxes will help but may not be sufficient

When direct and indirect subsidies are considered, gas heating is not the cheapest option. A carbon tax is needed to correct that problem. With such a correction, electricity becomes the best option almost everywhere, unless it is produced by coal.

However, when replacing oil or gas heating by electricity, the cheapest conversion is to install baseboard units. This is more efficient than gas but not the best option. In a warm climate, air-to-air heat pumps probably have the best long-term performance. In a cold climate, geothermal heat pumps are necessary to maintain high performance during very cold weather.

In any climate, the best options require a significant up-front investment that will later allow major reductions in energy consumption. Because of these investment requirements, consumers may not choose the lowest-cost option in the long-term.

4.9 Doomed, unless we implement large programs to retrofit buildings and convert them to electricity

We should not forget another major issue related to heating or cooling buildings: many of them are deficient from an efficiency perspective. They often have poor quality windows and doors, or low insulation in walls and ceilings. In warm climates, many buildings do not have insulation at all. Many countries do not understand the benefits of insulation when it is very hot, to keep a house cool, and to reduce air conditioning needs.

Neglecting (for one minute) the size of buildings, we can summarize the issues with three priorities:

- Implementing large retrofit programs to improve the efficiency of buildings.
- In warm or temperate climates, replace any fossil fuel combustion with air-to-air heat pumps.
- In cold weather, when possible, replace fossil fuel combustion with geothermal heat pumps.

Due to smaller energy invoices, these actions are generally cost effective, as the investment will be paid back in 5 to 10 years. Unfortunately, most households, commerce, and industries implement efficiency measures only if their financial payback is 2 to 4 years. To overcome choices based on short-term financial calculations, governments must implement major programs to ensure massive implementation of the most efficient options.

In cold weather, it is the case for geothermal systems. The underground portion of such a system requires a large investment, in the order of 25 000$. If the building is relatively big, this is economically justified. A good example would be a government program to convert all schools and public buildings from gas to geothermal heat pumps. All these buildings have space for the geothermal system, notably school yards or parking. Other markets, such as commercial or industrial buildings, could also be a priority.

And let's finish by repeating the most important issue: the need to control the total space requirements, for residential buildings, but also commercial and institutional buildings.

5

Transportation: The *Big Lie* concerning zero-emission vehicles

5.1 Key issues, answered by *political correctness*

- Electric cars are zero-emission.
- The cost of electric cars will continue to diminish.
- Hydrogen will reduce GHG emissions.

These statements are false or misleading.

5.2 The misleading *zero-emission* of electric vehicles

In the 1990s, there were many debates on electric cars, mainly in the United States. In that period, environmental groups were more vigilant than they are now. Understanding that most electricity came from fossil fuels, they said that *zero-emission* vehicles should be called "elsewhere-emission" vehicles. Now, this is *politically **incorrect***, even in regions where more than 80% of electricity comes from fossil fuels.

Many countries have made strong commitments to battery-powered electric vehicles. Let's be clear: for the long-term, we also support the development of electric vehicles, as it is the most efficient transportation technology. The alternative, hydrogen, is about three times less efficient (see next section).

However, considering the urgency of the climate issues, can electric vehicles greatly reduce greenhouse gas emissions in the next 20 years? With a life-cycle approach, we will see that the answer is clearly NO.

To put in perspective the performance of electric cars, let's first mention a few relevant statistics:

- In 2019, what percentage of national electricity came from coal? 20% in the United States, 60% in Australia, 54% in India, 65% in China[62].
- The largest producer of lithium batteries is China. In 2019, it was home to 73% of global lithium cell manufacturing capacity[63]. The U.S. was far behind at 12% of global total. In China, lithium production is supported by energy from coal.
- For comparison purposes, what are the direct GHG emissions from a conventional car? Typically, a car in Canada will travel 11 000 miles/year (18 000 km), and in the U.S. about 15 000 miles/year. For a compact car, the GHG emissions are **3 and 4 tonnes of CO_2/ year**.

---

[62] U.S. Energy Information Administration.
[63] Bloomberg NEF.

What are the greenhouse gases (GHGs) emitted in manufacturing the batteries of electric vehicles?

There is much debate on this issue. Promoters of electric cars assume that, in the future, all electricity will be clean, thereby ensuring low emissions from manufacturing. This assumption is unreasonable for the next 20 years.

Based on current production, mostly in China, the GHG emissions of lithium batteries are 150 kg of CO_2 /kWh[64]. For a Chevrolet Bolt, manufacturing the 60 kWh batteries emits **9 tonnes of CO_2**. Some electric vehicles have more batteries than this.

Do electric cars effectively replace conventional gasoline cars?

In 2017, a survey[65] of electric car owners in the U.S. was made. It revealed that nearly half the households that owned an electric car had 3, 4 or 5 vehicles. Most likely, these owners bought an electric car as an "addition", not to replace a conventional vehicle. Not considering the batteries, two studies conclude that building a car emits **7 tonnes or 10 tonnes of CO_2**[66].

---

[64] Science & Vie, 2018, Mia Romare, Lisbeth Dahllöf, The Life Cycle Energy Consumption and Greenhouse Gas Emissions from Lithium-Ion Batteries, IVL Swedish Environmental Research Institute, 2017.

[65] Clean Technica, 2017.

[66] CIRAIG study quoted in the magazine Protégez-vous, 2020.

What is the source of electricity used to fill the batteries?

In most countries, the dominant sources of electricity are coal or natural gas. How do these two sources of electricity affect the GHG emissions of an electric car traveling 15 000 miles /year (24 000 km)?

- If the electricity comes from gas, an electric car emits **2 tonnes of CO$_2$ per year**.
- If the electricity comes from coal, an electric car emits **4.2 tonnes /year**. This is higher than the 4 tonnes /year of a conventional car.

In many U.S. states or Canadian provinces, coal is still the main source of electricity. Even in these places, electric vehicles are officially called "Zero-Emission" vehicles.

Technical calculations:
Chevrolet Bolt consumption:
120 Miles per gallon equivalent or 5,7 km/kWh:
    24 000 km require 4.2 MWh
LCA data for electricity:
Electricity from coal: 1000 kg CO$_2$/MWh x 4.2 MWh = 4.2 t CO$_2$
Electricity from gas: 480 kg CO$_2$ /MWh x 4.2 MWh = 2 t CO$_2$

If hydropower is the main source of electricity, can electric vehicles be zero-emission?

In the province of Quebec, most electricity generation is from hydropower. Proponents of electric cars, therefore, assume that they are powered 100% by clean energy. Is this assumption valid? In Quebec, during a given year, the total amount of hydropower that can be generated is fixed, because it depends directly on the amount of rainfall. And every year, Hydro-Quebec exports a lot of electricity to the United States, reducing the use of natural gas there. Hydro-Quebec often boasts about the emissions avoided in the U.S. as a result of its exports.

This context means that each kWh consumed by an additional electric car is a kWh that can no longer be exported to the United States. Considering this fact, what is the effect of replacing a conventional car with an electric car in Quebec? It reduces GHG emissions in Quebec by 3 tonnes, but increases emissions in the US by 2 tonnes. This is far from *zero emission*.

If Quebec wants to claim that its electric cars are 100% "fueled" by hydropower, the province must launch hydropower projects, built specifically to power the future electric vehicle fleet. This is not the case, as development has now been halted.

For massive market penetration without subsidies, the cost of electric cars must be greatly reduced. Is that probable?

Like other new technology, electric vehicles can be the focus of unrealistic marketing. Since the year 2000, Hydro-Quebec and the province of Quebec have been promoting electric vehicles, notably because it has clean hydropower. In 2000, the Quebec government predicted that there would be, in the province, one million electric vehicles on the road in 2020. The reality was 37 000 electric vehicles, including hybrids.

In 2021, the true cost of an electric vehicle (without subsidy) was still nearly twice the cost of a similar conventional vehicle. This was due to the batteries, which cost about 40% of the total cost of an electric car.

The electric car lobby has copied the strategy of the wind power industry:

- First, they insist that the costs have been dramatically reduced. Official statistics show that batteries' cost has been reduced after 2010. However, a portion of these reductions was due to massive subsidies. A few examples: for its battery factory, Tesla received an incentive package worth 1.3$ Billion from the State of Nevada (2015)[67]; it also received a subsidy of 1.2$ Billion for a new factory in Great Britain (2020). China, a major producer of batteries, has constantly supported its industries with state funds.
- Secondly, the electric car lobby also predicts that the cost of electric cars will become so low that massive market penetration is unavoidable. Such a promise is not realistic: the official data on batteries show that, recently, the costs have stabilized. And producing more batteries will multiply the need for a specific group of metals[68]: Aluminum, Cobalt, Lithium, Manganese, and Nickel. Many factors point to an increase in the costs of these metals: larger consumption of metals creates economic scarcity, increasing costs; current mining practices are often

---

[67] Wikipedia, 2021.

[68] World Bank, The Growing Role of Minerals and Metals for a Low Carbon Future, 2017.

dangerous for workers and will need to be improved; the best mining sites are the first to be exploited, and future sites will require larger investments per unit of production. All these factors will increase the cost of batteries and electric vehicles. Some of these metals are needed for wind power **and** electric cars, creating scarcity and higher prices.

In Quebec in 2020, only 7% of new cars were electric or rechargeable hybrids. At the global level, sales of electric vehicles were still small, despite subsidies to each buyer:

- 13 000$ in Quebec;
- 2500$ to 7500$ in Germany depending on battery size;
- 3000€ to 5000€ in France;
- 5000$ to 8500$ in China (2016) even if 65% of electricity comes from coal.

What is the probable future behavior of electric car owners?

The efficiency of an electric motor is about four times greater than an internal combustion engine. As a result, the user costs of an electric vehicle will be much lower than those of conventional cars.

In economics, it is essential to distinguish between the fixed costs of a vehicle (notably the buying price) and its variable costs (notably the fuel). When a product has a high fixed cost, it tends to be used intensely to justify the original expense. When the variable costs are high, this tends to reduce the use of the product, as the cost of each use is perceived by the consumer.

Compared to conventional vehicles, electric vehicles have much higher fixed costs and much lower variable costs. This cost structure is a very strong incentive to drive more. The fact that users believe that they do not pollute at all (false) is an additional incentive. So it is probable that electric vehicles will increase suburban sprawl.

These factors allow an obvious conclusion: zero-emission cars do not exist; it is a marketing strategy by large corporations who want to sell cars.

For the long-term, electric vehicles may represent the future; but they will not make serious GHG reductions in the next 20 years. Their contribution to reducing emissions could be further delayed if fossil fuels are not quickly eliminated from electricity generation.

To control climate change, wasteful energy consumption is also an issue. And recent trends are worrying, as electrification is becoming a justification for wasteful behavior. There is much marketing of electric motorized recreation, such as personal watercraft, all-terrain vehicles, or snowmobiles.

A worst-case of wasteful behavior is illustrated by one article[69] in *Popular Mechanics*. It describes the future of electric pick-up trucks, with prices over 70 000$:

- A Chevrolet Hummer with motors of 1000 HP, going 0-60 mph in 3 seconds.
- The huge Rivian R1T with a large towing capacity and 180 kWh of batteries; their manufacturing will emit 27 tonnes of CO_2. This is equivalent to a small car fueled by gasoline, traveling 210 000 km (130 000 miles).

Electric vehicles, without other strong policies, can become the justification for wasteful consumption.

---

[69] Popular Mechanics, July 2020.

5.3 The excessive promises about Hydrogen

To be used as fuel, Hydrogen must be in the form of H_2. On planet Earth, there is no significant source of H_2. Currently, industries who want to use H_2 in their process must spend a large quantity of energy by reforming natural gas from CH_4 to H_2. With this method, using H_2 as a fuel would emit twice CO_2 than burning the natural gas directly.

Another method to produce H_2 is by splitting water (H_2O) through a process called electrolysis. **For the two production methods, the energy required is much greater than the energy available in the final H_2.** Hydrogen is not a source of energy, only a tool to store or distribute energy. It is not in competition with fossil fuels but with batteries and electric transmission lines.

Despite this reality check, promises about the future of Hydrogen have been excessive for the last 40 years. In 1987[70], an article in the Washington Post describes a debate in Congress, where promoters promise that the U.S. is about to enter the *Hydrogen society.*

In 2002, a similar debate happened in Congress on the miracles of Hydrogen and fuel cells (that consume H_2 to produce electricity). As a result, President Bush[71] announced a $1.2 billion research initiative:

---

[70] Washington Post, Fueling the Future with Hydrogen, 6 Sept. 1987.

[71] January 2003 State of the Union address, described by Joseph J. Romm, The Hype about Hydrogen, Issues in Science and Technology, spring 2004.

> *so that the first car driven by a child born today could be powered by hydrogen, and pollution-free.*

The magazine *Wired*[72] then proclaimed, *How Hydrogen Can Save America.* In August 2003, General Motors said that the promise of Hydrogen cars justified delaying fuel-efficiency regulations.

Life-cycle assessment can show the real potential of various transportation options. For the three options in the next table, we assume that the starting point is **100** units of clean electricity. We then estimate how much of this electricity serves to move a vehicle.

Promoters of Hydrogen insist that their option can be *zero-emission* if the electricity comes from wind or solar power. But, **compared to a battery-powered electric car, the hydrogen car needs three times more electricity** (see next table).

---

[72] Wired, April 2003.

Option	Life-cycle efficiency	Details of estimate
Tram or Light Rail Transit	87%	• 100 units of clean electricity go to a tram by transmission lines: loss of 8% • The electric motor is efficient at 95%: loss of 5%
Battery powered electric car	79%	• 100 units go to the charging station: loss of 8% • Batteries charging cycle: losses of 10% • Electric motor efficiency of 95%: loss of 5%
Hydrogen (H_2)	28%	• 100 units go to the H_2 factory by transmission lines: loss of 6% • Electrolysis factory extracts H_2 from water at 70% efficiency: loss of 30% • Pumping/compression of H_2 efficient at 85%: loss of 15% • In a car, the H_2 is consumed by a fuel cell with efficiency of 50%: loss of 50%.

In **2020,** Hydrogen was back on the list of options, receiving public funds. Why? Because large companies are promoting it for their benefit. For example, Boeing[73] is saying that it will replace oil in aviation.

---

[73] Boeing Environmental Sustainability Strategy, 2020.

Airbus makes a similar assertion but admits to major obstacles, notably concerning the H_2 storage on commercial aircrafts:

- Compressed H_2 is not an option because the tanks would be too heavy.
- Liquid Hydrogen is the only option.

H_2 is a gas at normal temperature. To liquefy H_2, it must be cooled at minus 253°C. It needs to be kept at that temperature on board the plane. This is a technical challenge that will itself consume large amounts of energy.

5.4 The misleading greenness of ethanol: case closed

The 1980s saw much marketing in favor of biofuels such as ethanol. There were debates on its real performance. In the U.S. and Canada, promoters gained national rules and gasoline distributors were forced to include 5% of ethanol in gasoline. This was the result of massive lobbying by the corn industry, as ethanol production creates a huge market for corn. Since then, many life-cycle assessments (LCAs) of ethanol have been made. There are disagreements among the LCAs: ethanol from corn could allow for a 30% reduction in GHGs or none at all. But even with a 30% reduction, the energy inefficiency seems unacceptable: the corn production and ethanol plant require seven energy units of fossil fuels, to produce ten energy units of ethanol. And it uses a large portion of the best agricultural land, driving up the prices of corn, an essential cereal.

More recently in Europe, biodiesel created a similar debate, due to rapeseed oil. The LCAs arrive at similar conclusions as in the case of corn ethanol. And because diesel engines are very polluting, most European countries are phasing out policies that previously favored diesel vehicles.

5.5 Doomed, if subsidies to electric vehicles are a substitute for real actions

The technological development of batteries and electric vehicles must continue. It will be a serious option in 20 years when electricity is much cleaner. In the short-term, governments should eliminate subsidy programs for car buyers: these programs do not lead to significant emission reductions and encourage sprawl.

We must understand that politicians have adopted electric car subsidies to compensate for their lack of courage in managing transportation. In the next chapter, we will look at more effective and courageous options.

6

Transportation: The large potential of trams

6.1 False impressions due to *political correctness*

- Diesel buses greatly reduce GHG emissions.
- The underground metro is the best public transit mode.

These statements are false or misleading.

6.2 Misleading perceptions about diesel buses

The table[74] on the following page presents life-cycle data typical of Canadian conditions.

It shows the performance of diesel buses is dependent on load factor. In suburban conditions (modest load), a small car has a better performance than diesel buses.

---

[74] Table based on 10 international life-cycle assessments quoted in Luc Gagnon, Jean-François Lefebvre, *Test climat : Réseau express métropolitain (REM)*, Syndicat canadien de la fonction publique, Coalition climat Montréal, mai 2018.

In urban conditions, diesel buses have acceptable performances with intensive use. Montreal is a good example. But even in urban conditions, a car with two persons on board has a better performance.

Life-cycle GHG comparaisons

Detailed table in *Test-climat du REM*

Options	Load factor	g CO2 eq. / passenger /km	
		Energy source	Life-cycle
Medium-size car	1 p /car	Gasoline	300
Small hybrid car			200
Urban bus (STM)	**High**		**150**
Urban bus	Medium	Diesel hybrid	200
Suburban bus	**Low**		**270**
Suburban train	Low		110
Skytrain (REM)	Medium		60
Suburban train	Medium		30
Tramway	**Medium**	Hydropower	**20**
Tramway	**High**		**15**
Trolleybus	Medium		30
Metro (first 30 yrs)	Medium		70
Metro (first 30 yrs)	High		40
Metro (after 30 yrs)	High		10

Diesel hybrid buses reduce emissions, only if customers are crammed in

6.3 A serious option: direct electrification of trains, light rail, trams, and trolleybuses

Because of climate change and other issues, households with numerous cars will need to change their lifestyle and probably take public transit. But politicians, to be *politically correct*, do not want to mention this obvious fact. As a result, they support the electrification of individual vehicles, which require batteries. There is a more efficient technology to electrify transportation for linear services: **direct overhead lines**. These can include

- Trolleybuses or trams replacing diesel buses in urban neighborhoods;
- Electrification of suburban trains;
- Electrification of freight trains.

In most cases, such technologies replace diesel engines (with efficiencies of about 20%) with electric motors (with efficiencies of about 95%); they allow major reductions of GHGs and many other pollutants. Moreover, they do not need batteries, avoiding the charging cycle which reduces efficiency. Batteries are also very heavy: to understand this issue, we can compare a battery-powered electric bus with a trolleybus. In the electric bus, the weight of batteries is 3 to 4 tonnes, equivalent to constantly transporting 40 to 50 passengers.

Another efficient option can seriously reduce emissions: intercity electric trains replacing short journeys by commercial aircraft. Fast trains going from one downtown to another one can replace longer commercial flights. This is a realistic option, as shown by France, who already has many TGV trains (*trains à grande vitesse*) between cities.

These electrification options could seriously reduce GHG emissions. But they do not necessarily address the issue of suburban sprawl. This issue is discussed in the next section.

6.4 The best public transit modes to reduce sprawl

In many countries, the zoning laws are not strong enough to stop sprawl. It is the case in Canada and the U.S. The site and density of developments are directly dependent on transportation infrastructures. The first essential action is simple: zero extension of roads or highways.

New infrastructure must become a strong incentive to increase density within cities. Choosing the proper public transit mode is the key to achieving this.

The table on the next page is a summary of transit choices. We avoid using the common expression *Light rail transit (LRT)*. It often creates confusion, as it can refer to 4 different modes: commuter train, tram, skytrain (similar to a tram but on an aerial structure), and streetcar (small tram).

Before making detailed comparisons of modes, here are a few public transit issues:

- The table shows the wide range of capacity among modes. It is essential to choose "The right mode in the right place". The choice depends on the expected ridership. If the mode capacity is too great, public money is wasted. If the mode capacity is too small, the service will be poor, with passengers crammed in peak periods.

Transit modes	Mode (Example)	Passengers /train	Typical line; no. of trips /day
	Metro Montreal	1500 Montreal	150 000 to 300 000
	Skytrain Vancouver, Montreal	600	80 000 to 100 000
	Commuter train (Montreal)	1200	10 000 to 35 000
	Trams (More than 350 cities)	192 (27 m) to 662 (90 m)	10 000 to 100 000
	Trolley bus (Vancouver)	120	10 000 to 20 000
	Articulated bus: diesel or hybrid	105	10 000 to 20 000
	Bus: diesel or hybrid	75	10 000 to 15 000
	Bus: Battery powered electric	55	7 000 to 10 000

- Public transit is used massively when many citizens live near efficient infrastructures and can **walk to a station**. Feasibility studies often focus on the number of citizens within a radius of 500m of a station.
- Each metro or tram station becomes a major attraction. Near a station, developers can promise future residents that they will benefit from good quality transit. They can also promise businesses that their workers will easily come to work.
- The choice of transit mode must not indirectly facilitate private cars' use. A metro or *skytrain* line can improve the flow of road traffic by removing hundreds of buses from the roads.
- Large parking at transit stations may increase ridership but prevents high-density development around stations. Parking is normally used by commuters and supports suburban sprawl.

How to choose between an underground metro and trams on the surface?

Despite their costs, politicians often promise metro extensions, for one reason: they do not want to disturb car owners, as metro lines reduce road congestion, facilitating the use of cars. In contrast, trams use one or two lanes on normal roads; this is seen as a drawback by some politicians. But within a strategy to reduce the number of cars and GHG emissions, this drawback becomes an advantage, as it reduces the space available for cars.

Despite this warning, some metro lines are justified. Most large cities include some high-density neighborhoods and high-density business centers. Metro lines are justified when traffic is high, at least 100 000 trips/day. The expected ridership is a good indicator that such an infrastructure is needed.

A metro line, however, is about ten times more expensive than a tram line, per station (or mile). Based on the principle of "The right mode in the proper place", trams are better adapted when the density is not very high. Expressed another way, it is possible, with the same investment, to build ten tram stations, compared to one metro station. These stations will attract many more users because they can walk to one of a great number of stations.

Trams can greatly reduce the economic costs of public transit. Their wide range of capacity means that the choice of the tram can be closely related to the expected ridership. The right mode should be well adapted to the density of neighborhoods.

How to choose between trams and buses?

Buses are more flexible than trams, to make detours and serve low-density neighborhoods. The bus is, therefore, the logical choice for networks of about 10 000 trips per day. But diesel buses are noisy and polluting; they are often obstacles to development. A good indicator is the expected price of a property: the presence of a bus stop in front of a home will reduce its value.

In contrast with buses, trams tend to attract development and increase the density of cities. Here are the reasons:

- Their network is more visible and accepted as permanent.
- They are less polluting and quieter.
- They are more reliable, especially in the snow.
- In a hot climate, they can more easily be air-conditioned.
- In dense neighborhoods, one tram can replace 3 to 5 buses, greatly reducing the noise and congestion.
- Wheelchairs entering a bus create a long delay. In trams, they can enter without delay.

Trams can significantly reduce the economic costs of public transit, compared to buses. The use of electricity is four times more efficient than diesel, greatly reducing energy costs. And depending on ridership, one tram unit can replace 2 to 5 buses. This means much lower operating costs and fewer drivers.

Many studies have documented the effect of trams on development (see summary on the next page). They confirm that the metro and the tramway attract residential and commercial development, unlike bus systems. The tramway often has the best economic performance because, for a given investment, it is possible to install ten tram stations against a single metro station.

Findings from a comparison of 60 cities[75] Jeffrey Kenworthy, *Why Rail Systems Are Essential in Creating Eco-Cities*	
Tramway and metro stations have a great influence, concentrating development	• Rail can be very powerful in influencing the form and scale of development • Rail can attract major re-development over a 15-20 year period
Tramway and metro systems increase the use of public transit, but not buses alone	• Strong rail cities have more public transport passenger boardings than weak rail and no rail cities • Strong rail cities capture more than a four times greater proportion of overall motorized passengers on transit than no rail cities
Parking around stations should not be encouraged	Excessive emphasis on parking, including Park & Ride, destroys the urban design and civic qualities of sub-centers
General conclusion: metro and tramway are essential to reduce car dependency	Urban Rail Systems are the key to the renaissance in public transport worldwide and a key to reducing automobile dependence. We will not change any significant size city into a more ecological model without high-quality urban rail

---

[75] Jeffrey Kenworthy, Eco-city: Ten Key Transport and Planning Dimensions for Sustainable City Development, Environment and Urbanization, April 2006.

Findings of the Victoria Transport Policy Institute[76], Canada	
The bus is harmful if we want to attract development	Findings suggest that increasing regular bus frequencies results in lower house values for properties located in the vicinity of regular routes
Tram or metro stations increase the value of surrounding properties	*"the value of properties along tram network increased by more than 40%"*
Tram and metro attract more users than buses	Public transit trips increased an average of nearly 16% in Rail & Bus cities but only 1.7% in Bus-Only cities

The benefits of trams are demonstrated in hundreds of cities, notably in Europe. These cities have supported their dense neighborhoods with trams, improving their quality of life. The city of Lyon is a convincing case: many years ago, the city chose to develop metro stations, which did not affect the use of private cars. It then decided to implement numerous tram lines, with significant reductions in the use of cars.

How to choose between trams, and battery-powered electric buses?

Some politicians assume that the tramway is no longer useful, as it is easier to introduce battery-powered electric buses on a large scale. This presumption is unfounded for two reasons:

---

[76] Todd Litman, Victoria Transport Policy Institute, Evaluating Public Transit Benefits and Costs; Best Practices Guidebook, 27 October 2019.

- A battery-powered electric bus must carry 3 or 4 tonnes of batteries, a weight equivalent to 50 passengers at all times. It, therefore, has less capacity than a diesel bus, and costs twice as much.
- And above all, for major networks, one tram set can replace 2 to 5 buses, with significant reductions in operating costs.

Do battery-powered electric buses have a future? Yes. They will be the best option for small networks with a ridership of fewer than 10 000 trips per day.

6.5 How billions of dollars of transit can support suburban sprawl and the use of private cars

We use the case of Montreal to show how bad transit choices can serve to maintain the dominance of cars. Since 2017, a private promoter has proposed to build the *Réseau Express Métropolitain* (REM). Here are a few parameters that describe the project[77]:

- To ensure that the project does not disturb the flow of private cars, the promoters chose a *skytrain* that will always be aerial or underground. Because it is automated, it cannot roll even one mile on regular roads.
- The project will be entirely built by a private company, the CDPQ-Infra which will invest about

---

[77] Luc Gagnon, Jean-François Lefebvre, *Test climat : Réseau express métropolitain (REM)*, Syndicat canadien de la fonction publique, Coalition climat Montréal, mai 2018.

$3.6 billion. To cover that investment and to exploit the service, the CDPQ-Infra will charge the Quebec government about $600 million /year.

- The total building cost of the REM is $10 billion when government contributions are included: about $6.5 billion. Governments are basically "giving" many infrastructures to the CDPQ-Infra: the Mount Royal tunnel, the Two Mountains train line, and two lanes of the new Champlain Bridge.
- The REM replaces two existing services with high ridership: the Two Mountains commuter line and, on the Champlain bridge, the busway serving the South shore of Montreal. The REM expects to serve 160 000 trips per workday (or 80 000 users). But because it is mainly replacing existing lines, the $10 billion will only convince 10 000 car drivers to adopt transit. This is $1 million per car. The REM is even less effective than this, because most of these 10 000 car drivers will still use a car to go and park at the stations.
- The REM is designed to serve far-away suburbs. To please suburban drivers, many stations will include free or subsidized parking. On the South Shore, one new station will serve a large suburban shopping center.

After the project launch (2016-17), official forums were available to express opposition to the project. Environmental and transit groups had many reasons to oppose the project: privatization of public transit, additional sprawl, choice of the wrong mode, a huge waste of money without any serious reduction of the number of cars. But, because of *political correctness*, they did not dare criticize the project simply because it is transit. It shows how *political correctness* can have negative effects, indirectly maintaining the dominance of private cars.

And even after such a bad experience, the Quebec government continued to make the wrong choices. As the REM was being built, an extension of the Blue line of the metro was authorized. It is a short line with five stations, costing about $6 billion. This is $1.2 billion per station, probably the most expensive transit project in the world, per station. This metro extension will replace a bus network that currently carries **30 000** passengers /day. Car owners will be happy to get rid of that bus network.

6.6 Doomed, unless we implement mega-developments of tram lines

Tram lines multiply the number of transit users, reduce the use of private cars, and reduce suburban sprawl by drawing development near stations.

The case of Montreal can illustrate the potential of this option. With the $10 billion, the REM adds 12 new stations. With another $5 billion, the Blue metro line adds five stations. With these budgets, it would have been possible to implement seven tram lines with about 150 stations. This would have made Montreal the world's capital of public transit, with increased urban density and fewer cars. Instead of that, the current domination of private cars will persist.

7

Transportation: Changing the "rules of the game"

7.1 False perceptions, in agreement with *political correctness*

- Road congestion must be reduced because it increases GHG emissions.
- There is no need to change the current system, as new efficient technologies will solve the problems.
- Within the overall climate strategy, we must not exaggerate the importance of transportation, as it is responsible for only 25% of GHG emissions.

These statements are false or misleading.

7.2 The overall responsibility of road transportation

In official statistics from different countries, transportation represents 25 to 35% of GHG emissions. These statistics are misleading, because they include only GHGs that come out directly from vehicle exhausts. They do not include emissions from the following activities: manufacturing of vehicles; oil extraction and treatment in refineries; oil delivery to gas stations; cement and metals to build roads and bridges; HFC losses from car air conditioning equipment; extraction of all the related resources, etc. Many of these activities are energy-intensive. Once we consider the overall transportation system, the responsibility can reach 50% of GHG emissions.

7.3 The misleading effects of congestion

At first glance, road congestion reduces the speed of cars, thereby increasing fuel consumption. Many politicians, to be *politically correct*, promise to reduce congestion. They say this to please drivers. But this is a small portion of the overall picture.

As described in chapter 3, suburban sprawl is the main factor making us dependent on numerous cars. Suburban sprawl also tends to lengthen the average trip by car. **In the current system, with free roads, what is the only factor that slows suburban sprawl? Road congestion.** A quote by David Owen[78] sums it well:

> *One of the few forces with a proven ability to slow the growth of suburban sprawl has been the ultimately finite tolerance of commuters for long, annoying commutes.*

____

[78] David Owen, The Conundrum, p.84.

Local politicians in the suburbs are very much aware of this limitation; they are constantly asking for more new roads and bridges, to allow further development of their suburb.

Road congestion is acceptable because it reduces energy consumption and GHG emissions. Newman and Kenworthy[79] demonstrated that the most congested cities have the lowest energy consumption (in transportation). Congestion is also a very fair "policy", as the people who suffer from congestion are generally the people who cause it.

7.4 The misleading effects of more efficient cars

Because of technological improvements, car owners can now buy cars that consume less fuel than 30 years ago, without sacrificing comfort or convenience. Here are a few potential consequences of more efficient cars:

- A driver does not change his habits, drives the same distance, and saves money on less fuel. But this means he has more money to spend on other goods that may also require a lot of energy.
- Another driver now has the perception that fuel is less expensive because he consumes less fuel per mile. He will drive more.
- Another example is the case of drivers who buy more expensive and efficient hybrid vehicles. Why spend more? Because they have the intention or the need to drive long distances.

---

[79] Newman and Kenworthy, Cities and Automobile Dependence, 1991.

- Because of technological improvements, people can buy a Sports Utility Vehicle with more comfort or convenience than a car, without a serious increase in fuel consumption. The efficiency gain serves comfort instead of reduced consumption.

These reactions are components of the **rebound effect**. This is NOT to say that smaller and more efficient cars are not desirable. They are needed, but we must understand that about 50% of the expected benefits are lost to the rebound effect.

Globally, to reap 100% of the efficiency gains of new technology, it is necessary to increase the fuel or transportation cost (for example, with a mileage tax).

Another relevant question: how is the efficiency gain achieved? If smaller cars are sold instead of large vehicles, there may be significant gains. But in the past decades, one of the tools to improve efficiency in the U.S. was the Corporate Average Fuel Efficiency standard, where each manufacturer must achieve an average fuel efficiency for all its cars. To respect this standard, the car builders do not sell more small cars; they create lighter cars by using more aluminum and more plastics in building them. In many cases, the energy required to build cars has increased, as aluminum is more energy-intensive. So the standard reduces the official consumption per mile but increases the energy consumption in other portions of the life-cycle.

7.5 Fewer roads and parking spaces

Even with electric cars, cities need to reduce the number of vehicles massively. Experience has shown that the number of cars always increases with additional road and parking spaces. Consequently, cities must offer fewer road and parking spaces to reduce the number of cars.

Building tram networks contribute in two ways: they take away space previously used by cars; the tram stations attract development, often on parking lots.

The number one issue, however, is economic: for drivers, the cost of roads is zero or very small; and often, the cost of parking is zero.

7.6 Essential tools: road tolls, large taxes on fuel and parking lots

A few environmental groups are proposing surcharges on the purchase price of gas-guzzlers. This strategy has a weakness: purchasing a vehicle is a fixed cost. Once this expense is made, the user sometimes says to himself: "After making this big expense, I might as well use my vehicle, and avoid paying for public transportation". This illustrates a limitation of electric vehicles: they are extremely expensive to purchase (fixed cost) and inexpensive to operate (variable cost). Once purchased, there will be a strong incentive to use them.

These comments lead to an obvious conclusion: to reduce the use of vehicles; it is necessary to increase their variable costs, not their fixed costs[80]. Here are examples of effective tools: (see next table for the summary).

- Tolls are very effective because the rate can be modulated by time of day, thereby reducing some of the social costs of congestion.
- A consistent objection to tolls or parking fees is the difficulty of collecting the fee. Many opponents claim that tolls slow down traffic. This claim is false with modern technology. For example, in Norway and Australia, nearly **all vehicles are equipped with a transponder,** which allows direct communication with any toll booth without slowing

---

[80] Luc Gagnon, Jean-François Lefebvre. Jonathan Théorêt, *Modalités et avantages d'une réforme fiscale écologique pour le Québec : Mythes, réalités, scénarios et obstacles,* 2014, 71 p. Work done for the Commission d'examen sur la fiscalité du Québec.

down. A transponder costs about $50 per vehicle. This technology is also appropriate for parking fees, implemented by municipalities or private businesses.

- Among current practices, free parking and monthly parking passes should be eliminated. After a driver has paid the monthly fee, there is no incentive to take public transit.

GHG emissions management tools for transportation

Actions	Effects
Subsidies to fossil fuels	Very harmful
Subsidies to ensure that roads are free	
Subsidies to ensure that parking is free	
Income tax	No direct effect
Sales tax on all products	
Vehicle efficiency standards	Small positive incentive
Vehicle registration fees	
Carbon tax	Strong positive incentive, but small effect on electric cars
Fuel tax	
Mileage tax (or kilometer-based tax)	Strong positive incentive
Road tolls	
Hourly parking fees	

The level of transportation fees must be significant. A few years ago, the Quebec government was proud to announce its carbon tax as a tool to reduce GHG emissions. Since the implementation of the tax, the strong growth in the number of cars and SUV has continued. Some may say that the tool is not efficient. The issue is simply the size of the tax. In 2021, for the average car owner, it was equivalent to

$300 per year. So implementing the tax was like saying the following sentence to the average car owner: "each year, we used to subsidize you by $3600; in the future, the subsidy will be only $3300". No wonder it had so little impact.

To understand the need for large fees, we can quote the studies[81] of Donald Shoup: he concluded, in 2005, that off-street free parking in the U.S. represents a subsidy of about $3.75 per gallon of gasoline ($1.00 per liter). This value would be much higher now.

Some studies show that, in the long-term, parking fees or constraints have a strong effect, convincing drivers to use public transit or carpool.

7.7 A tax on aviation fuel

Another important source of emissions is air travel. For the next 20 years, we cannot expect any technological progress that will significantly reduce GHG emissions from commercial aircraft. The only option is **fewer flights**.

The best tool to achieve this is an international tax on aviation fuel. Is it a problem if some countries refuse to implement such a tax? Not really, because participating countries can simply impose a surcharge on flights arriving from recalcitrant countries.

---

[81] Donald Shoup, The high Cost of Free Parking, 2005.

7.8 Doomed, unless we reduce subsidies to transportation

As mentioned previously, the technological development of electric cars should continue; however countries must cancel the massive subsidy programs on purchase price. These subsidies do not reduce GHG emissions significantly and are an incentive to sprawl.

To reduce emissions from transportation, two very efficient actions must be implemented quickly:

- In cities, developing numerous tram lines to reduce the use of cars and fight suburban sprawl. This is not necessarily difficult for politicians, as such development creates jobs and economic activities.
- **The most important action is adopting economic incentives to cover the huge subsidies for transportation.** The potential tools: fuel or carbon taxes, mileage tax, road tolls, and parking fees. It is up to environmental groups or public transit advocates to make proposals in favor of such green taxes; they may respond that this is an unpopular proposal. But if environmental groups do nothing on this issue, who will? (In chapter 10, we will discuss how to make green taxes politically acceptable).

8

Food and Agriculture: One action for the climate

8.1 Conflicting goals and *political correctness*

There is an abundance of recommendations on how food should be produced and consumed by citizens. They come from different perspectives:

- Health: Organic production is justified by the need to reduce the use of pesticides and protect health. Moreover, organic production is dependent on organic fertilizer, instead of synthetic fertilizer from fossil fuels. As a result, groups assume that organic products do not emit greenhouse gases (GHGs).
- Local economy: Many groups insist on the importance of buying local products to help local farmers and reduce delivery distance. As a result, they assume that buying local products cause a significant reduction of GHGs.
- Waste management: Environmental groups insist on the need to reduce packaging, notably plastics produced from fossil fuels. They assume that this allows for a major reduction of GHGs.

We can see that concerns are very diverse: health, economic benefits, waste management, avoiding fossil fuels, and plastic. Unfortunately, if the goal is reducing GHG emissions, these recommendations are inefficient.

8.2 Misleading perceptions about organic production

Synthetic fertilizers have a bad reputation because they are produced from natural gas. There is, however, one important question: Can synthetic fertilizers be massively replaced by organic fertilizers? On a planet of 8 billion people now, to 11 billion by 2100, it is not realistic to assume that we can massively abandon synthetic fertilizers, without reducing the food supply. Even if we could replace them, a large portion of organic fertilizers would come from livestock, which are a major source of GHG emissions.

Other misconceptions are related to an assumption that agriculture can be *zero-emission*. Organic farming still requires mechanized activities. Even organic fertilizer will decompose, emitting carbon dioxide and methane. Organic farming is a great method for producing healthy food, but it is a significant source of GHG emissions. And several important cereal production will emit GHGs. For example, rice production cannot be *zero-emission*, because it must be grown on flooded soils: inevitably, this causes methane emissions.

8.3 Misleading perceptions about local food consumption

Many groups insist on the need to buy local products. This is a good practice to support local farmers. But a detailed analysis cannot conclude that this is a priority. The reason is simple: within the life cycle of most food, the energy in food distribution represent a small portion of energy spent. Many life-cycle assessments arrive at about 5%.

Packaging generally represents another 5% of food total carbon footprint[82]. So at least 80% of emissions are caused by growing and preparing the food.

One recent study concludes that transportation can be 15% of emissions; but this study included upstream transportation of resources (such as fertilizers). Buying local produce does not eliminate these life-cycle transportation requirements, as many agricultural inputs need to be delivered to a farm before growing food.

8.4 Misleading perceptions about packaging

There is much debate on the impacts of plastics. We will discuss this in the next chapter. We do not dispute the fact that some packaging is excessive. But relative to food, LCAs are quite conclusive: the largest energy consumption is in food production, not in packaging. And one priority in reducing energy consumption is to reduce the spoilage of

---

[82] "Taking a closer look at paper cups for coffee", VTT Technical Research Centre of Finland, 2018-19.

produce. In some cases, plastic packaging helps to achieve this.

K-cups of coffee can illustrate the reality of packaging. Some environmental groups have identified this as a worst-case of over-packaging. One LCA has assessed the overall GHG impacts of coffee production and distribution. For coffee consumed as K-cups, only 3% of impacts came from producing the K-cups[83]. The one major issue was coffee production. If a consumer, to avoid K-cups, uses a filter coffee machine and makes a double dose of coffee to avoid starting again, he may waste coffee and multiply his GHG emissions.

8.5 Fact: meat consumption by people is the number one problem

Meat production is responsible for high GHG emissions in various ways: large quantities of cereals are required to feed the cattle; beef and cows emit, through their digestive system, significant amounts of methane; manure from pork and poultry emits methane; meat must be refrigerated in all steps of processing and distribution.

A research group at Montreal University (CIRAIG) made a complete picture of GHG emissions of a household due to food consumption: the two main sources of GHGs are Meat and fish at 36% and Dairy products at 15%.

---

[83] Quantis Canada, 2015, quoted in magazine Protégez-vous, 2020.

Food category	kg CO_2 eq. /kg food
Beef	26
Lamb	23
Cheese	8
Pork	7
Chicken	5
Fresh fish	4
Milk	2
Beans	1

This table summarizes studies on GHG emissions from food production, compiled[84] by the Food and Agriculture Organisation. All data allows one conclusion: when a consumer decides to eat beef instead of cereals or plant-based protein, he can multiply his GHG emissions by a factor of 10.

The CIRAIG[85] (2020) also made a picture of GHG emissions, in Quebec, due to food consumption: Again "Meat, fish, and dairy products" cause 51% of GHG emissions and food packaging 5%.

We did not find studies indicating how humans can produce good quality food without significant GHG emissions. **So again, the notion of *zero-emission* does not exist**.

8.6 Fact: meat consumption by pets is the number two problem

Food consumption in each household can cause significant GHG emissions. And in household assessments, some "members" are not included: cats and dogs. Having pets is very *politically correct*; few groups dare to discuss their consumption of meat that cause large GHG emissions.

---

[84] FAO data quoted in New Scientist, Feb. 16, 2019, p. 33.
[85] CIRAIG, quoted in the magazine Protégez-vous, 2020.

UCLA professor Gregory Okin[86] has calculated the meat consumption of pets in the U.S.: For a human population of 330 million, there are 163 million dogs and cats, eating 25 to 30% of national meat consumption. This meat consumption is responsible for 64 million tons of CO_2 per year, equivalent to the emissions of **15 million cars**.

Some would say that these emissions can be easily reduced, because dogs can live without much meat. In reality, pet food marketing often includes statements such as "with real meat".

U.S. cats and dogs consume more meat than all the meat consumed by the 200 million citizens of South East Asia

---

[86] Gregory S. Okin, Environmental impacts of food consumption by dogs and cats, PLoS One, Aug. 2017.

(Thailand + Vietnam + Cambodia + Laos). Because of *political correctness*, environmental groups rarely raise this issue.

Dogs also contribute to another collective mistake: choosing a house in a faraway suburb with a large lot. Households often make this choice to have space for their dogs. The energy implications of such a choice are rarely considered.

As with other climate issues, the goal here is not to blame individuals who have pets. The goal is to understand the real issues that we must address collectively. **We cannot progress if we cannot identify the real issues.**

One collective tool that can reduce the impact of pets is simple: adopt a regulation to limit the quantity of meat in dog or cat food. Nutritionists could define a maximum percentage of meat in cat or dog food. It would be different for cats and dogs.

8.7 Should we implement a tax on meat consumption?

Implementing a tax on any food is politically difficult. The United Kingdom has raised many controversies, by implementing specific taxes, notably on sweet drinks, for public health purposes.

In previous chapters on energy, housing, or transportation, we showed the importance of implementing carbon taxes. What is the effect of a universal carbon tax on meat consumption?

In 2020, the Grain Farmers of Ontario opposed the Canadian carbon tax. This opposition was not entirely rational, because it neglected the fact that the Canadian carbon tax is implemented with a direct credit to households. Sylvain Charlebois, an expert on agriculture, clarifies what this means at the level of Canadian consumers[87].

> *Consumers will get rebates. Depending on where you live, rebates could exceed $500 annually. Lower-income, and often food-insecure households, with lesser carbon footprints, will receive a rebate equal to, and perhaps more than, the increase in the cost of goods with the new tax. This may be good news for a demographic in need of cash, but in the end, we are all likely to see inflation go up, which will eventually hit the prices we find at restaurants and in grocery stores, whether we get rebates or not. Energy is required to get the food we need, but the extent of its impact on food prices remains unknown. One report published in 2012 suggested that the effect of a $50 per tonne carbon tax on food prices would be 3%.*

The key issue is how this average 3% increase will affect different foods. When a citizen consumes beef instead of cereals, he is multiplying his cereal consumption and related energy inputs, **by a factor of ten**. Moreover, beef production emits large amounts of methane, a powerful greenhouse

---

[87] Sylvain Charlebois, Carbon Tax Will Increase Grocery Prices in Canada: Expert, April 3, 2019.

gas. As a result, the price increase will be very different for beef and cereals: basically, the 3% overall increase means an increase of 1% for cereal consumption and probably 30% for beef. **A carbon tax will effectively put pressure to lower the consumption of meat**. A specific tax on meat production or consumption is **not** necessary.

9

The *small actions* culture, a distraction

9.1   Small actions justified by *political correctness*

Relative to environmental protection, many groups insist that *every small action is important*. Here are some of their recommended actions:

- Recycling of residential waste.
- Reducing the use of plastic bags.
- Replacing old light bulbs with LED lights.
- Avoid buying products online, because it generates deliveries.

Many people assume that these individual actions can make a difference. We will now look at them from a life-cycle perspective.

9.2 Misleading perceptions about plastic

Environmental groups have been critical about the use of plastic while proposing the following strategies.

Banning the sale of plastic water bottles?

- With a focus on plastic pollution in the oceans, many groups in rich countries campaign on this issue. They ask for a ban on single-use plastic bottles. In reality, there is practically no relation between ocean pollution and the use of plastic bottles in rich countries; nearly all plastics ending up in oceans come from rivers located within developing nations (8 of 10 rivers located in Asia). A New Scientist article[88] describes the source of the problem:

  There are 2 billion people in this world who don't have proper waste collection systems.

- To significantly reduce the plastic pollution in oceans, rich countries should provide financial aid to developing nations, to help them manage waste properly. And we also need to consider substitutes for plastic bottles[89]:

  You must use a steel water bottle 500 times for its carbon footprint to shrink to less than of a disposable PET bottle.

- Instead of banning plastic bottles, maybe it is more efficient to simply charge 25¢ per bottle as an incentive to reduce their use.

---

[88] New Scientist, How to solve a problem like plastics, May 19, 2018, p.27.

[89] New Scientist, May 19, 2018, p.31.

Replacing single-use containers by reusable glass or ceramic containers?

- Glass is an option to replace plastic in many food services. But glass is much heavier than plastic and will increase energy consumption in delivery. When the glass container is reused, the assessment must consider the energy required to bring it back and wash it perfectly.
- Energy spent on transporting and washing glass can be significant. This is demonstrated by LCAs that compare glass bottles and aluminum cans for beer. If the aluminum is recycled, there are no significant differences in GHG emissions, because of the extra weight and washing of glass.
- Another research[90] compared single-use paper cups and ceramic cups. It concludes on the issue of washing:

The major factor affecting the climate impact of ceramic cups was the efficiency of washing. This includes the use of clean water, detergent, energy and the waste water treatment. The study found that dishwashing causes more than 90% of the life cycle emissions of a reusable cup. The study suggests that even when washed efficiently, ceramic cups need to be used at least 350 times before having a smaller carbon footprint than that of a paper cup.

---

[90] "Taking a closer look at paper cups for coffee", VTT Technical Research Centre of Finland, 2018-19.

Replacing single-use plastic bags with reusable bags in grocery stores?

- Most opponents to plastic bags call them "single-use". From the start, this is unfair, as many single-use bags are reused, either for waste or another purpose.
- Depending on different technology and life-cycle assessments, producing one large reusable plastic bag (HDPE) emits about 20 to 50 times more GHGs than the production of "single-use" plastic bags of similar capacity. The HDPE bag must be reused many times to break even. In the long-term, reusable bags probably have a better performance. A study[91], done for the government of Australia, concludes that reusable Green Bags have a better performance than single-use plastic bags: they allow a reduction of 6 kilograms of GHG emissions per household per year. In comparison, the cars of a typical household can emit more than 6 000 kilograms /year (2 cars at more than 3 tonnes each).
- And the 6 kilograms reduction is uncertain because the study assumed that each reusable bag is effectively reused constantly. In reality, people sometimes forget to bring their reusable bags and buy some more: as a consequence, many households end up with about 15 reusable plastic bags at home, with 10 of them rarely used.
- Other options may be worse[92]:

---

[91] Hyder Consulting, Life Cycle Assessment of Supermarket Carrier Bags, Environment Agency of Australia.

[92] New Scientist, How to solve a problem like plastics, May 19, 2018, p.31.

A cotton tote bag must be used 131 times before its environmental cost falls below that of a disposable plastic bag, mostly because of the impact of growing cotton.

- Maybe, it is more efficient to simply charge 25¢ per single-use plastic bag as an incentive to reduce their use.

9.3 Recycling of unsorted recyclable products?

Numerous publications describe the theoretical energy benefits of recycling. They show the efficiency of a manufacturing plant that uses virgin material compared to recycled material. Some examples: producing tin cans from the minerals requires three times more energy than from recycled tin cans; producing PET bottles from the products of oil refineries requires four times more energy than from recycled PET bottles. This data at the manufacturing plant is misleading. The challenge is getting clean tin cans to the plant without other metals. The challenge is getting the PET bottles together without the other six types of plastics that can sabotage the process.

There are debates on the choice between sorting waste in a large recycling plant, or at home (or in a commerce). Sorting at home may be better if households do it perfectly. But this is not always the case, so some level of sorting is needed anyway at the final plant. One article[93] concludes that there is no obvious winner:

----

93 New Scientist, Waste not...?, July 22, 2017, p. 40-43.

> *But the more complex the household sorting task becomes, the more likely householders are to give up and simply pitch something into the rubbish. As a result of this trade-off, local authorities often lump all recycling into a single bin...*

Many cities have decided to do the sorting at a large plant. But the collection of items is energy intensive: in a residential or commercial setting, a big garbage truck, fueled by diesel, must stop one hundred times (or more) to gather one truckload of various recyclable products. It must then deliver this to a big recycling plant far away from neighborhoods.

A portion of these products are not recyclable or not recycled (ex. colored glass, many types of plastics, etc.). Another portion was recyclable but not anymore because of contamination (ex. cardboard with food stains). Another portion is not recyclable because mixed with another product (ex. aluminum and plastic in tetra pack foil).

In many countries, some products are not recycled because there is no industrial demand. This is often the case for recycled glass with small pieces of different colors. After sorting in a plant, recycled products are often of poor quality, and industries are not ready to buy them. The consequence: they are exported to developing nations, notably China, which are less concerned about quality. This means additional energy spent on transportation.

Overall, yes, recycling does save energy but much less than expected when we consider the complete life cycle. And most of the benefits of recycling come from a few products, such as metals.

In Europe, many countries burn waste to produce energy. This means that plastic has value because it burns well. But in North America, incinerators are often judged unacceptable. This means that plastic recycling has very little value because few plants make an effort to separate the **seven** different types of plastics.

Recycling is also affected by urban density. For example, on many street corners in Barcelona, there are large bins allowing pre-sorted collection of products: one bin for glass, another for paper/cardboard, another for metals, etc. This is efficient because of the high population density: citizens go to the bins on foot, and when a truck stops, it can gather the wastes of hundreds of households. In the early days of recycling, such a system was set up in low-density suburbs, to allow citizens to pre-sort of products. But households were frequently making an extra trip by car to recycle a few items. In this case, the energy consumption of cars was probably greater than the energy gains of recycling.

So all this discussion leads to one simple fact: recycling is a good practice to reduce the need for landfills but it does not significantly reduce energy consumption.

9.4 Misleading perceptions about online shopping

Some groups say that delivery services provided by Amazon are environmentally destructive. A Canadian group specializing in life-cycle assessment has looked at shopping habits, comparing buying on-line or going to a shopping center. They conclude that there are no significant differences because both habits involve a lot of driving.

The number one issue is the quantity of times products are bought. One consumer mistake is buying multiplied objects, knowing they can return many of them. Either on-line or in-person, this multiplies the energy consumption.

9.5 Environmental education censored by *political correctness*

A few authors have been courageous and concluded on the major issues, even if they are not *politically correct*. One study is so relevant to this book that we include the complete abstract[94]:

> *Current anthropogenic climate change is the result of greenhouse gas accumulation in the atmosphere, which records the aggregation of billions of individual decisions. Here we consider a broad range of individual lifestyle choices and calculate their potential to reduce greenhouse gas emissions in developed countries, based on 148 scenarios from 39 sources. We recommend four widely applicable high-impact (i.e. low emissions) actions with the potential to contribute to systemic change*

---

[94] Seth Wynes, University of British Columbia, Kimberly A Nicholas, Lund University, The climate mitigation gap: education and government recommendations miss the most effective individual actions, Environmental Research Letters, January 5, 2017.

and substantially reduce annual personal emissions: having one fewer child (an average for developed countries of 58.6 tonnes CO_2-equivalent (t CO_2e) emission reductions per year), living car-free (2.4 t CO_2e saved per year), avoiding airplane travel (1.6 t CO_2e saved per roundtrip transatlantic flight) and eating a plant-based diet (0.8 t CO2e saved per year). These actions have much greater potential to reduce emissions than commonly promoted strategies like comprehensive recycling (four times less effective than a plant-based diet) or changing household lightbulbs (eight times less). Though adolescents poised to establish lifelong patterns are an important target group for promoting high-impact actions, we find that ten high school science textbooks from Canada largely fail to mention these actions (they account for 4% of their recommended actions), instead focusing on incremental changes with much smaller potential emissions reductions. Government resources on climate change from the EU, USA, Canada, and Australia also focus recommendations on lower-impact actions. We conclude that there are opportunities to improve existing educational and communication structures to promote the most effective emission-reduction strategies and close this mitigation gap.

Their recommendation concerning the number of children is controversial; it is discussed in chapter 11 on population growth.

9.6 Overall profile of a household

Other studies confirm what activities cause the most GHGs. This table shows the picture of a typical household in Quebec. It was produced by a university group specialized in life-cycle assessment[95].

Emissions per Household	Tonnes CO_2 eq
Daily transportation	16.2
Food	6.9
House	6.3
Clothes	3.9
Other	8.4
Total yearly	41.7

The four major sources represent 80% of household emissions.

9.7 A few individual actions can make a difference, but they face an "uphill battle"

At the level of each household, what are the actions that can seriously reduce their emissions? Having no car, or only one car and using it less; using public transit or a bicycle regularly; choosing a smaller habitat in an urban neighborhood instead of a large house in a suburb. In sum, choosing a habitat that minimizes motorized transportation and energy consumption. However, as described in the

---

95 CIRAIG quoted in Protégez-vous, May 2020, p. 15.

previous chapters, **the current economic and fiscal system greatly supports NOT choosing these actions.**

Avoiding air travel is also significant. Even if LCAs arrive at different results for commercial flights, the GHG emissions are always substantial: Montreal to Paris round trip for 1 person: 1.2 tonne[96]; a transatlantic flight round trip for 1 person: 1.6 tonne[97]. Currently, most countries provide subsidies to airports and do not include aviation fuel within their commitments to reduce emissions. And socially, taking a distant vacation is often considered essential.

The overall conclusion: for any significant contribution, a household must make difficult choices against social trends. The slogan *every little action counts* is a distraction from what is needed.

Eating less meat is possibly the only recent positive trend.

9.8 Doomed, unless individual action is aimed at collective changes

Individual action to reduce the energy consumption of one household is better than nothing. But what we need are changes that will seriously reduce the energy consumption of a great number of households.

With this in mind, one crucial action is getting involved in politics:

---

[96] CIRAIG, Canada.
[97] Wynes and Nicholas, 2017.

- Voting for the proper political candidate that understands climate change and the needed changes;
- Even more efficient is political action in favor of collective measures, such as a large carbon tax.

Major progress is only possible by creating incentives for many people; collective action must change the "rules of the game".

10

Green taxes or Ecological fiscal reform?

10.1 Major changes required in fiscal and budget policies

In the previous chapters, we constantly showed that GHG reductions depend on major changes in fiscal rules and public budget priorities. The following list is a summary of these potential changes:

- In electricity generation, it is essential to replace coal and gas, by wind power, hydropower, and nuclear energy. To achieve this, a minimum carbon tax of 135\$ per tonne of CO_2 is needed.
- Suburban sprawl must be stopped. One essential policy is stopping subsidies to new highways or transit designed to favor suburban development.
- Governments must limit tax exemptions for large houses, in faraway suburbs.
- Heat pumps must replace gas in heating. To incite this, a carbon tax is needed on fossil fuels used in buildings.

- To reduce oil consumption and the average length of car-trips, we need many policies: stopping subsidies to oil extraction and pipelines; implementing a mileage tax or increasing taxes on fuel.
- In cities, to reduce the number of cars and favor public transit, we should charge the real cost of parking.
- To reduce the energy consumption in transportation, including from electric cars, an **energy** tax is the best option. Subsidies to electric cars should be stopped.
- To reduce the number of flights, implement a carbon tax on aviation fuel.
- In many countries, municipal taxes are based on the total value of a property (land + building). The tax rate should be much higher on land than on buildings.
- To reduce the use of single-use plastic, we suggested a 25¢ tax on each bottle or bag.

Although these measures are needed, **we do not propose that politicians or states implement this list of fiscal changes**. A long list of separate policies will be perceived as an attempt to increase the overall tax burden. The good news is that there is another method.

10.2 Doomed, unless we implement ecological fiscal reforms to overcome *political correctness*

When green taxes are needed, the primary reaction is defined by *political correctness*: groups and politicians simply avoid the topic. Unfortunately, politicians do have reasons to be afraid. Experience has shown that opponents to green taxes can manipulate the debate and say that any green tax is a serious increase in the overall tax burden.

This is why we need an *ecological fiscal reform*, which is quite different than individual green taxes[98]. **Within an ecological fiscal reform, politicians can guarantee that various tax reductions will compensate for any increase in new taxes.**

Such an *ecological fiscal reform* can be summarized in four components:

a) Identification of the current subsidies that increase energy consumption and GHG emissions. This is not difficult, as even the International Energy Agency has constantly identified the problem, asking countries to stop subsidies to fossil fuels. Abolishing these subsidies means that large public funds will be available to facilitate an ecological fiscal reform. Moreover, this may greatly simplify the fiscal system.

b) Identification of practices, within the current tax structure, that may be unfair or inefficient. There are many obvious cases: income tax applied to low earnings; excessive income tax levels that discourage citizens to work; high universal sales tax that require much bureaucracy; sales tax on socially desirable services, such as insurance premiums. Within the reform, the government proposes to **greatly reduce** these taxes. We cannot be too specific on

---

[98] Luc Gagnon, Jean-François Lefebvre. Jonathan Théorêt, Modalités et avantages d'une réforme fiscale écologique pour le Québec : Mythes, réalités, scénarios et obstacles, 2014, 71 p. Work done for the Commission d'examen sur la fiscalité du Québec.

these options because they vary between states and countries.

c) The reform **implements one or two major tax increases** to avoid a long list of controversial green taxes. The major changes could be either a carbon tax or an energy tax applied to ALL energy sources, including aviation fuel. To simplify the message and the implementation, the best option is probably an energy tax, modulated to be higher for fossil fuels than for wind power, hydropower, and nuclear energy. Overall, it is probable that the new fiscal system will be more simple, fair and efficient than the old system. Politically, this method appears to be the only way to implement a carbon or energy tax that is **large enough** to have a serious impact.

d) The fiscal and budgetary reform is then implemented with a specific law to ensure that tax reductions are **implemented at the same time** as the additional energy or carbon tax. Past experiences have shown that citizens do not have confidence in promises about future tax reductions.

There are examples of such reform, even if they are limited and incomplete. The province of British Columbia implemented a modest *ecological fiscal reform* in 2008. The implementation of a carbon tax was compensated by various tax reductions guaranteed within one law. In 2018, the Canadian government adopted a similar approach for its carbon tax. In these two cases, the only problem was the size of the tax: a modest tax level has a modest effect on emissions.

An *Ecological fiscal reform* could have a structural effect on all activities that consume a significant amount of energy. Households, industries, and cities would all implement measures to reduce the taxes paid, reducing GHG emissions. It is similar to implementing a long list of individual green taxes (previously discussed).

Even if such a reform would be very efficient, a few additional efforts may be needed to attenuate the forces causing suburban sprawl:

- Revision of the municipal fiscal system to reward urban density;
- Revision of utilities invoicing practices that currently penalize high-density and reward low-density development.

10.3 Fiscal green tools: negative, neutral, and positive effects

Within an *ecological fiscal reform*, a few guidelines can help define the priorities.

Harmful policies, relative to energy consumption

Fiscal /Budgetary policies	Probable effects
Fossil fuel subsidies	Large subsidies keep prices low, leading to overconsumption
Subsidies to expand and maintain free roads	Free roads encourage • solo driving, an energy-intensive option • trucking, which is less efficient than rail and ship
Subsidies to suburban developments	Sprawl multiplies trip length and increases car dependency
Corporate /municipal subsidies for free parking	Free parking is a key factor in choosing car over transit

Neutral policies, relative to energy consumption

Fiscal /Budgetary policies	Probable effects
Income tax	High level has negative effects on desire to work more
Corporate payroll contributions	Employer's contributions may be an obstacle to hiring permanent employees
Uniform sales tax on all goods and services	Sales tax does not distinguish between polluting products or socially desirable services

Weak Policies, relative to energy consumption

Command & control / budgetary policies	Probable small effects, with occasional perverse effects
Vehicle efficiency standards	• Improved efficiency per mile without reducing ownership or use rates • The rebound effect eliminates a portion of the benefits
At car purchase, *bonus* on efficient vehicles and *fee* on gas-guzzlers	• Improved vehicle efficiency • The *bonus* may increase the rate of vehicle ownership
Monthly or yearly parking fee	Once parking is paid for, there is no incentive to choose another mode

Strong policies to reduce energy consumption

Fiscal /budgetary policies	Reduction of consumption, if the tax is significant
Carbon tax	Very efficient because it puts pressure on: 1. Energy source choice 2. Energy efficiency 3. Overall energy consumption
Tradable permit system with permits sold at auction	• Similar as a carbon tax • Much less efficient if many permits are given freely by governments
Energy tax	Efficient, because it puts pressure on: 1 Energy efficiency 2. Overall energy consumption

Fuel tax	• Similar to carbon tax for transportation. • Electric cars avoid this tax (even if electricity comes from fossil fuels)
Mileage tax	• Efficient to cover all road use • Small cars may pay the same as large car • Electric cars must pay this tax
Road tolls	• Decrease in vehicle utilization rates • Possibility of targeted regional pricing • Time-differentiated pricing to reduce congestion
Daily or hourly parking pricing	• Strong incentive for public transit or carpooling • Decrease in vehicle utilization rate

10.4 Overall benefits of an *Ecological fiscal reform*

If the energy or carbon tax is high, the reform will cause large reductions in GHG emissions. There are other benefits:

- By affecting the choices of economic actors, the reform will create a strong incentive for more efficiency measures. This can reduce the expenses of many industries or households.
- Reducing sprawl can have economic benefits for everybody by reducing public expenditures for linear infrastructures such as roads, bridges, aqueducts, sewers, and streetlights. It can also reduce expenses for numerous other services, such as deliveries.
- Households will become wealthier if they spend less on transportation.

- Finally, the reform can easily protect the poorest segment of society: for example, a credit can be provided to eliminate any negative economic impact.

Using such economic tools tends to favor GHG reductions that are the least expensive for society. It is not the case for the current approach: governments often adopt complex legislation for minimal reductions.

10.5 Social acceptability: prohibitions or incentives?

Many left-wing groups do recognize that reducing GHG emissions is essential and urgent. Some are even ready to prohibit many non-essential motorized recreation activities, such as snowmobile, motocross or private watercraft. The huge impacts of climate change could justify such actions. This however, poses many ethical dilemma. Who should decide what is acceptable? What is essential? What restrictions are sufficient?

From an ethical perspective, economic instruments are preferable to prohibitions. When governments use economic instruments, it does not mean that households or industries do not make reductions; they will have to choose among numerous actions to reduce emissions.

For example, after an *ecological fiscal reform*, a household will adapt by choosing a few of these actions:

- Using public transit or a bicycle more often;
- Selling one car or buying a more efficient car;

- Traveling less often; choosing a closer vacation destination;
- Traveling more by train, less by air;
- Chose services or restaurants that are closer to home;
- Abandon some motorized recreation; adopt active recreation;
- Try to work more often from home (if the employer agrees);
- When choosing the next house, consider transit service and car dependence;
- Adopt energy efficiency measures at home.
- Eating more cereals and less meat.

The list could be longer. **The point here is that each household will decide.** If they decide that air travel is important, they can cut their energy consumption in other areas and continue flying. Within the current economic and budgetary traditions, the government chooses what is better, even if the choices are unfair and inefficient. One example are subsidies to rich households that can afford an electric car.

A long list of actions is also available to industries. To make a profit, they will need to minimize the energy or carbon tax they pay. Often, they will choose the most efficient actions to reduce emissions.

What is surprising is that right-wing parties or governments often express their opposition to green taxes. This is an emotional and irrational response. Using economic instruments allows more individual and corporate freedom, which is among their priorities. And the rational actions of citizens and corporations will then allow emission reductions without additional involvement by governments.

11

Population growth, the forbidden issue

The population issue demonstrates, again, the need to distinguish between:

- Individual or household behavior.
- Collective problems, which can only be solved by various states or institutions.

11.1 The population issue, as extreme *political correctness*

There is consensus that the current global level of consumption is unsustainable. For many resources such as fisheries, forests, or agricultural soil, the current rate of exploitation greatly exceeds their renewal capacity.

In 2020, this global overconsumption was done by 7.7 billion people. International institutions predict that the population will probably grow to 11 billion in 2100. This growth will make it extremely difficult for many countries to reduce GHG emissions. International institutions do not mention this evidence, even the Intergovernmental Panel on Climate Change. This is a collective failure.

This failure was summarized in 2018 by the group *Population Connection*[99]:

> *By failing even to mention population growth in the thirty-three-page "Summary for Policymakers" of its new publication, Global Warming of 1.5°C, the Intergovernmental Panel on Climate Change (IPCC) offered less than the whole truth...*
>
> *The IPCC "Summary for Policymakers" finds room to call for ecosystem-based adaptation, ecosystem restoration, biodiversity management, sustainable aquaculture, efficient irrigation, social safety nets, disaster risk management, green infrastructure, sustainable land use, and water management. But not one single word about population stabilization.*
>
> *Ignoring the impacts of soaring population growth on climate change is like failing to mention the Himalayas while describing Nepal.*
>
> *Overpopulation remains the elephant in the room. Why the silence? The hard truth is that many experts worry they might offend someone, somewhere. This is a shameful betrayal of reason.*

---

[99] Population Connection Internet site, John Seager, IPCC must end its silence on population, 2018.

Discussing population growth is *politically **incorrect***. At the individual level, we understand that addressing the number of children per household is ethically delicate. But women must endure social pressure to have more children, and international institutions should at least consider the collective dimensions of this social pressure. Here are a **few examples** of social "rules" supporting large families:

- We are an ethnic minority. If we do not increase our birth rate, our economic conditions will worsen.
- We are an ethnic majority. If we do not increase our birth rate, we may become a minority.
- We are a religious minority. If we do not increase our birth rate, we will be oppressed by other religions.
- We are a language minority. If we do not increase our birth rate, our language will disappear.
- The population in our (rich) country is getting old; if we do not increase our birth rate, we will experience a shortage of young workers.

Accepting such social rules will make climate change unstoppable. Even without considering climate change, these rules create obstacles to development, for everybody, including minorities. And the justifications are often made at the expense of women's rights. Few groups or countries will acknowledge these collective issues.

11.2 Population and the lessons of Life Cycle Assessments

Many citizens or groups assume that it is possible to live without any significant impact on the planet. Life-cycle assessments (LCAs) show that this is a big illusion. Many environmental impacts are created even when citizens have an "exemplary" lifestyle. And even if burning fossil fuels is drastically reduced, there are many other sources of GHGs, not related to fossil fuel combustion:

- Rice paddies and many other agricultural activities emit methane and CO_2.
- Production of biofuels (ethanol or biodiesel) from crops or wood.
- Using wood as fuel emits large quantities of CO_2: it is one option to reduce the use of fossil fuels.
- The petrochemical industry, with or without fossil fuels, will still be a source of GHG emissions.
- There is no substitute for cement, which is responsible for high GHG emissions.

The population and climate challenges are different in rich and developing countries. In the next two sections, we will try to summarize them, by answering three questions:

a) Can we significantly reduce GHG emissions by applying policies aimed at "consuming **better**" (electric cars instead of conventional cars)?
b) Can we significantly reduce GHG emissions by applying policies aimed at "consuming **less**" (fewer cars, less travel…)?
c) Once effective strategies to reduce GHG reductions are implemented, is it essential to stop population growth?

11.3 Population growth and GHG emissions in developing countries

In African nations, the population growth rate varies from 2.2 to 4% per year. This growth is explained by an average rate of 4 or 5 children per woman. **With a growth rate of 4%, a country multiplies its population by three, every 28 years**. This is an obstacle to improving the quality of life. Despite this, many environmental groups say that population growth in developing countries is not a problem because of their low *per capita* emissions. Does this make sense? We can answer by looking at potential actions to control or reduce GHG emissions.

a) **In developing countries, can we significantly reduce GHG emissions by applying policies aimed at "consuming better"?**

Many small actions, proposed in rich countries, are not relevant, for example consuming fewer plastic bags. Another potential action would be replacing conventional cars by electric cars. Is it reasonable to assume that developing nations can develop wide networks of charging stations and that citizens will be able to pay $50 000 for a small car? Another option would be to eat less meat and more cereals, but meat consumption is already low. In fact, trends show that when countries get richer, they increase their meat consumption. So globally, it is difficult to assume any large GHG reduction with these strategies.

b) In developing countries, can we significantly reduce GHG emissions by applying policies aimed at "consuming less"?

In poor countries, the *per capita* emissions are very low because most citizens are not rich enough to buy many consumer goods. Currently, in the world, one billion people do not have access to reliable electricity supply. If new wind power is installed, it will probably serve to increase supply, not to replace fossil fuels. Many households live without essential products such as a fridge. Air conditioning is reserved for minorities or public buildings. When households can afford a car, it is often a very small or used car.

In rich countries, energy consumption *per capita* can be ten times greater than those of African countries. The approach of "consuming less" is appropriate for rich countries, not for poor countries or households.

c) In developing nations, once an effective strategy to reduce GHG reductions is implemented, is it still essential to stop population growth?

Even if the current energy consumption is low, one trend is probable for the future: **most citizens of developing countries aspire to lifestyles similar to those in rich countries**. This will increase their energy consumption. Moreover, many citizens will want or will need to migrate. Their preferred destinations will be rich countries with large *per capita* consumption. As a result, these migrants will probably increase their energy consumption and GHG emissions.

Managing climate change needs a long-term commitment. One obvious conclusion: **the most critical contribution that developing countries can make is stopping their population growth**. International institutions do not recognize this fact.

11.4 Population growth and GHG emissions in rich countries

a) In rich countries, can we significantly reduce GHG emissions by applying policies aimed at "consuming better"?

For households, it is possible to reduce GHG emissions by making better consumer choices. But five chapters of this book show that such reductions are small, relative to overall climate goals. For example, a household that chooses reusable plastic bags, instead of single-use plastic bags, can reduce their GHG emissions by 6 kilograms /year. But their two cars emit more than 6 000 kilograms /year in Canada, or 8 000 kilograms /year in the U.S.

Another action would be to eat less meat and more cereals; this action can be meaningful because meat consumption is often large and excessive for a healthy diet. So globally, it is possible to achieve significant GHG reductions with these strategies. But they will be in the order of 20 or 30%. This is clearly insufficient: as discussed previously, rich countries need to quickly reduce their emissions by 85%.

b) In rich countries, can we significantly reduce GHG emissions by applying policies aimed at "consuming less"?

In rich countries, energy consumption *per capita* is very high. The approach of "consuming less" is appropriate for rich nations or rich households. Unfortunately, the current fiscal and economic system rewards overconsumption. This is why an *Ecological fiscal reform*, discussed in chapter 8, is essential to reduce GHG emissions.

c) In rich nations, once an effective strategy to reduce GHG emissions is implemented, is it still essential to stop population growth?

The answer to this question is quantified by Wynes and Nicholas's previously quoted article[100]. For rich European countries, the authors have calculated that reducing the birth rate is the most efficient action to reduce GHG emissions:

> *We recommend four widely applicable high-impact (i.e. low emissions) actions with the potential to contribute to systemic change and substantially reduce annual personal emissions: having one fewer child (an average for developed countries of 58.6 tonnes CO_2-equivalent (tCO_2e) emission reductions per year), living car-free (2.4 tCO_2e saved per year), avoiding airplane travel (1.6 tCO_2e saved per*

---

[100] S. Wynes, K. Nicholas, The climate mitigation gap: education and government recommendations miss the most effective individual actions, Environmental Research Letters, January 5, 2017.

roundtrip transatlantic flight) and eating a plant-based diet (0.8 tCO$_2$e saved per year).

They also have made the assessment for the U.S., where *per capita* energy consumption is even greater:

> *Once they accounted for the generational effect – that every child a person has is likely to have children themselves- having one fewer child was the single most effective measure, saving **120 tonnes** of carbon dioxide a year for the average U.S. citizen. The next-biggest impact, living car-free, came in at **3 tonnes** of CO$_2$ a year for the average person in the U.S.*

Unfortunately, many traditional economists or demographers do not consider climate change. They insist that birth rates in rich countries are too low and will create economic problems. One example is a biased article by Cave, Bubola & Sang-Hun[101]: the authors associate low birth rate with catastrophic concepts, such as population "stagnation" or fertility "bust". They predict economic catastrophe for countries like Japan, South Korea, Sweden, and Germany. These are the wealthiest countries in the world. If these countries cannot manage population stabilization, who can?

---

[101] Cave, Bubola & Sang-Hun, "Long Slide Looms for World Population", May 22, 2021, NY Times.

11.5 Doomed, unless we accept that population growth is a collective problem

Population growth is another issue where individuals or households should NOT be blamed. Having children is a source of happiness. The point here is that:

- Countries, states, and institutions should stop implementing measures to increase their population.
- International institutions should not blame countries who do adopt population control measures (ex. China). They have.
- Conventional economists or demographers who advocate increased birth rate must be confronted by the reality of climate change.
- Institutions and groups concerned with climate change must show the impacts of population growth. This is notably the responsibility of the *Intergovernmental Panel on Climate Change.*

Conclusion

1992 RIO CONFERENCE
Commitment to stabilize
GHG emissions

NOW IS THE TIME FOR ACTION !

1997 KYOTO PROTOCOL
Rich countries commit to
minus 5%

WE DID NOT RESPECT THE RIO COMMITMENTS, BUT THIS TIME, WE WILL.

2009 COPENHAGEN COMMITMENTS
↓ 17% in 2020, relative to 2005
↓ 20-30% in 2020, relative to 1990

WE DID NOT SUCCEED WITH THE KYOTO PROTOCOL, BUT NOW WE KNOW HOW TO PROGRESS.

2015 PARIS AGREEMENT
Countries have not implemented the
actions needed to respect the
Copenhagen commitments

THIS TIME, WE ARE AIMING FOR MINUS 50% IN 2050.

2021 GLASGOW CONFERENCE OF
PARTIES (COP26)
· 7% reductions due to the pandemic
· Major increases with economic
recovery

NOW IS THE TIME FOR ACTION, TO IMPLEMENT THE GOAL OF ZERO NET EMISSIONS.

2030 ?
Probably too late to respect the Paris
Agreement

TO MOVE FASTER, MAYBE WE SHOULD HAVE REJECTED POLITICAL CORRECTNESS.

At least five times in the past 30 years, climate change management has been trapped in the following cycle:

- Scientific institutions make a moderate assessment of climate risks.
- Politicians react with commitments to reduce emissions.
- Countries do not implement the programs required to meet the commitments.
- At each cycle, countries announce more ambitious commitments.
- Politicians can become more ambitious (apparently) simply by adopting goals to be respected in 20 or 30 years, when they will not be in power anymore.
- Again, countries do not adopt the needed short-term actions.

One goal of this book is to try to break that circle.

This book shows that, because of *political correctness*, the impacts of climate change are greatly underestimated, while the benefits of new technologies are greatly overestimated. This is a recipe for disaster.

We hope that readers will look at the whole demonstration of this book, not just one chapter or a few sentences out-of-context. The demonstration is based on a logical flow where each chapter builds on the previous ones.

International institutions, countries, and many experts constantly underestimate the impacts of climate change. They neglect feedback mechanisms that can speed up impacts. The debate about stabilizing climate at +1.5°C or

+2°C was a complete waste of efforts. **We are on a path to +5°C, with the probable collapse of modern civilization.**

Because of the previous point, we need urgent action to prevent catastrophic climate change. We need to replace coal and gas-fired generation in the next 2 or 3 decades. We must also replace gas and oil heating by clean electricity. Policymakers should understand that promoters of wind and solar power exaggerate the potential of their technology. These options can make a major contribution but, to be realistic, we also need large developments of hydropower and nuclear energy. High carbon taxes are essential to support such a development.

Because of the extensive impacts of climate change, we need to act on factors that cause a constant increase in energy demand. One major factor is the low-density suburban sprawl that can multiply the energy consumption of households. It can multiply the need for linear infrastructures by a factor of ten or more: roads, sewers, aqueducts, electric and gas networks, streetlights, etc. It makes each household dependent on numerous cars and multiplies the average distance traveled. Sprawl makes it nearly impossible to provide good-quality public transit. Because of *political correctness*, very few groups raise the issue of suburban sprawl. Exceptionally, they do speak about "urban" sprawl, even if the urban portion of cities is not sprawling. They avoid the word "suburban". Typically in Canada, governments subsidize each suburban household by at least 12 000$ **per year**, even more in the U.S. For many households receiving such subsidies, choosing the suburbs is a rational choice; we should not blame them for doing so.

A reduction of sprawl is only possible with a fundamental review of fiscal and budgetary policies.

Because of subsidies for suburban sprawl, the energy consumption of buildings has increased, despite some efficiency improvements. Different trends explain this energy consumption:

- Each new house is much bigger than 30 years ago;
- The number of people per household has diminished, creating demand for more houses;
- Mortgage fiscal exemptions have allowed households to buy larger homes and more secondary houses;
- Commercial buildings and houses have increased the number of appliances for refrigeration and air conditioning.

Actions are required to reverse these trends. And countries must implement mega-programs to improve the efficiency of buildings and electrify them with heat pumps.

Because of the speed and impacts of climate change, there is an urgent need to reduce GHG emissions from the transportation sector. In the last 40 years, the number of cars has dramatically increased because of suburban sprawl and massive subsidies for roads and parking. Economic incentives are necessary to compensate for these huge subsidies: carbon taxes or mileage tax, road tolls, parking fees, etc. To reduce the dependence on cars, policies must favor medium and high-density developments in urban neighborhoods. **Urban** light rail transit is one efficient tool to reduce the number of cars and increase density simultaneously. Governments should stop building new roads, as they increase sprawl.

A large portion of energy consumption and GHG emissions are related to suburban sprawl and transportation. There is one exception: meat consumption which multiplies impacts relative to the consumption of cereals. Carbon taxes will reduce meat consumption because of the energy consumption in meat production.

Instead of focusing on the main issues (sprawl, transportation or meat), many groups insist on *politically correct* actions: banning plastic bottles or "single-use" bags, recycling, buying organic food from local producers, etc. Even if these actions can provide various health and economic benefits, life-cycle assessments show that they do NOT significantly reduce GHG emissions.

Even if green taxes are not *politically correct*, they are essential to achieve any progress. One strategy can make green taxes socially acceptable: a broad *ecological fiscal reform*, with large reductions of income tax or sales tax, to compensate for the revenues of green taxes.

Life-cycle assessments show that every citizen, even with an exemplary lifestyle, will emit significant GHGs. The previous logical flow allows another conclusion: it is essential to stop population growth.

Another manifestation of *political correctness* concerns electric vehicles: this is a serious issue because it creates the illusion that we are effectively solving climate change. Currently, more than 80% of the electricity charging electric cars comes from fossil fuels. **The claim about *zero-emission* vehicles is currently a big lie**. If the electricity comes from coal, an electric car emits more GHGs than a conventional car; if it comes from gas, the GHG reduction is about 40%, far from *zero-emission*. Moreover, the pollution from battery production must be added to this assessment. The fact that a few **exceptional** places have clean electricity (Norway and three Canadian provinces) does not change this global assessment.

Promoters of electric cars promise that electricity and battery production will soon become clean. Are these promises reasonable? We can apply the previous logical flow to answer this question:

- Reducing GHG emissions is urgent. Options that reduce emissions in 20 years can be useful but not a priority.

- In electricity generation, the promises related to wind power are exaggerated. Replacing coal and gas will be very difficult. It will take a few decades.
- For the next 20 years, a large portion of the electricity that serves electric cars will come from fossil fuels.
- For the next 20 years, a large portion of the electricity that supports battery manufacturing will come from coal. China, the largest battery manufacturer, still produces most of its electricity with coal.
- As a result, electric cars will not greatly reduce GHG emissions in the next 20 years.

In 20 or 30 years, electric cars will become an important component of a climate strategy. Continuing technological development is desirable. In the short term, however, subsidies designed to increase commercial sales are very inefficient in terms of dollars per avoided CO_2.

The overall conclusion is simple and worrying: **if we do not eliminate *political correctness* from climate policy, our modern civilization is doomed.** Trying to be optimistic, if we do eliminate *political correctness*, there are many options to seriously reduce GHG emissions.

We must accept the lessons of the past 30 years: the escalation of empty promises does not lead to significant progress. We must move to real urgent action, such as thousands of projects based on wind, hydro and nuclear energy. Large programs are needed to eliminate the use of fossil fuels in buildings. Moreover, the transformation of taxation rules and public budgets is essential. For this, *ecological fiscal reforms* appear very effective.

Appendix: Greenhouse Gases

Greenhouse Gases (GHGs)		Main sources from human activities	Global* Warming Potential	Emission** Estimates 2019 Gt CO_2 eq.
Carbon dioxide	CO_2	-Coal, oil, natural gas, cement -Deforestation, land-use change	1	38 6.6
Methane	CH_4	Gas production /distribution, coal and oil production, livestock, rice production	28 to 34	11
Nitrous Oxide	N_2O	Agriculture (Synthetic fertilizers and manure)	265 to 298	2.7
Sulfur Hexafluoride	SF_6	Electricity distribution	23 500	
Hydrofluoro-carbons	HFCs	-Many industrial uses -HFC134 losses from automobile air conditioning units	12 to 14 800 1 100	1.4
Chlorofluoro-carbons	CFCs	New production banned		
Perfluoro-carbons	PFCs	Aluminum production	7 390 to 12 200	
Tropospheric Ozone	O_3	Due to NO_x and COV from road traffic		Not emitted directly

*The Global Warming Potential (GWP) is an index measuring the radiative forcing of a given substance (per unit mass), over a chosen time horizon. It is relative to the reference substance, carbon dioxide (CO_2). In the table, GWPs are estimates over 100 years. Over 20 years, the GWP of methane is 80.

Emission estimates[102] from IPCC AR6 Technical Summary. Emissions are expressed in Gt CO_2 **equivalent, which means that the different GHGs are added, after multiplying them by their GWP.

---

[102] IPCC Technical Summary. In Climate Change 2021: The Physical Science Basis. Contribution of Working Group I to the Sixth Assessment Report of the Intergovernmental Panel on Climate Change, Cambridge University Press, Cambridge, United Kingdom and New York, NY, USA, pp. 33–144.